BE GREEN HEROES

EDITION 2

BE GREEN HEROES

EDITION 2

Copyright © 2024 Sae Chang

All rights reserved

No part of this book may be reproduced, or stored in a retrieval system, or transmitted in any form or by any means, electronic, mechanical, photocopying, recording, or otherwise, without express written permission of the publisher.

ASIN-Kindle Edition: B0DHLRFTVW

ISBN-Paper Cover: 9798345635339

ISBN-Hard Cover: 9798344917191

Sae Chang, **BE GREEN HEROES**: FIGHTING WIDESPREAD MISINFORMATION AND PURSUING ENVIRONMENTAL LITERACY, HANDBOOK OF TEXTILE SUSTAINABILITY

Edition 2 Notes:

This second edition has been thoughtfully refined to enhance readability for a broad audience, welcoming both casual readers and industry professionals. Passages offering in-depth insights for industry professionals and students in related fields are clearly marked as ***Industry Specifics***. These sections provide valuable technical detail for those who seek it, while allowing casual readers to bypass them without losing the flow or central emphasis of each topic.

BE GREEN HEROES in Different Language Versions:

Italian

Titre: SIATE EROI VERDI: COMBATTIAMO CONTRO LA DISINFORMAZIONE, PERSEGUIAMO L'EDUCAZIONE AMBIENTALE, MANUALE DI SOSTENIBILITÀ TESSILE

Lien web et code QR:

https://amzn.to/4fBdbTG

French

Titre: BE GREEN HEROES: LUTTER CONTRE LA DÉSINFORMATION GÉNÉRALISÉE ET PROMOUVOIR LA LITTÉRATIE ENVIRONNEMENTALE, MANUEL DE DURABILITÉ TEXTILE

Lien web et code QR:

https://amzn.to/3UVzZ6M

Spanish

Título: SER HÉROES VERDES: LUCHA CONTRA LA DESINFORMACIÓN COMÚN Y LA BÚSQUEDA DE LA ALFABETIZACIÓN AMBIENTAL, MANUAL DE SOSTENIBILIDAD TEXTIL

Enlace web y código QR:

https://amzn.to/3H3vlv2

Table of Contents

Introduction

   About the Author

   About the Book

   Credits

Chapter I: Consumers, Be Aware and Beware!

   Section 1. Unfortunate Consumers

   Section 2. "Fast fashion"

   Section 3. Microplastics vs. Natural Fiber Dust

   Section 4. Unfortunate Disposal

   Section 5. Unfortunate Recycling

Chapter II: Basics

   Section 6. Overview - Fibers

   Section 7. Natural Fibers

   Section 8. Synthetic fibers

   Section 9. Semi-Synthetic (also known as Semi-Natural) Fibers

   Section 10. Manufacturing

Chapter III: Lifecycle Assessment

   Section 11. LCA - Farming of natural fibers / Fiber making of synthetic fibers

   Section 12. LCA – Manufacturing

   Section 13. LCA - Consumption

   Section 14. LCA – Disposal

   Section 15. LCA – Total Impacts

Chapter IV: Misinformation Crisis

    Section 16. Greenwashing

    Section 17. Self-Proclaimed Expertise

    Section 18. Organisations Lacking Expertise

    Section 19. Loopholes in Certification Programs

Chapter V: Solutions

    Section 20. United Efforts

    Section 21. Solutions

    Section 22. Clean Recycling Initiative™

    Section 23. BE GREEN HEROES

Closing Thoughts

Introduction

About the Author

My name is **Sae Chang**, and I am the founder of two organizations based in Montreal, Canada: **HEAT-MX™**, a developer of advanced thermal insulation materials, and **Clean Recycling Initiative™**, a non-profit organization dedicated to sustainable textile waste management.

With a background in **textile engineering** and hands-on experience across nearly every stage of textile manufacturing, from fiber production to garment assembly, I bring a comprehensive understanding of both the **technical** and **market-driven** sides of the industry. My career has included roles in product development, quality control, sales, and engineering, both within multinational corporations and through my own entrepreneurial efforts.

My journey has given me a unique ability to bridge the gap between **engineering** and **real-world market needs**, a gap that often leads to the failure of even well-intentioned sustainable products. This balanced perspective shapes the insights shared throughout *BE GREEN HEROES*, helping readers understand the deeper causes of environmental challenges in the textile industry, and how to approach solutions that are both practical and genuinely sustainable.

I also developed the **Clean Recycling Initiative™** technology platform, which reclaims textile waste with minimal environmental impact. Recognizing its potential for global good, I donated this technology to the non-profit under the same name, so it could be accessed without a financial burden by companies and the public worldwide.

Through this book, I aim to bring greater clarity to sustainability issues, challenge common misconceptions, and offer **science-based, actionable solutions**. I invite readers to engage with an open mind and a critical eye, as the road to genuine environmental progress requires both.

About the Book

Throughout this book, I will present many widely popular environmental initiatives and concepts, which are based on a significant lack of appropriate science, misleading the public. To encourage critical thinking, I propose several perspectives as a prelude, carefully selected to stimulate inquisitiveness in readers' minds.

- Widespread misinformation on the Internet
- Self-proclaimed experts educated by the Internet
- Organizations operating with a significant lack of the necessary expertise, knowledge and science
- Importance of science-based education

Importance of Comprehensive Views

In the realm of environmental management, it is crucial to maintain comprehensive views on the overall impacts of an action. For instance, while carbon emissions are undoubtedly a significant concern, they are not the only area that warrants attention. In my view, chemical toxicity is a critical issue that is largely overlooked in the current global sustainability efforts. Additionally, there are many other important factors that are being ignored. Focusing solely on a particular problem while neglecting potentially more impactful, hidden issues is analogous to only caring for one's skin without conducting necessary internal medical examinations. This neglect may lead to severe consequences that are more challenging to address in the future.

To some readers, my assertions may seem exaggerated. However, I can assure you that they are not. I will highlight numerous pressing issues addressing this concern with specific examples throughout the book. Many of them are closely related to our daily lives, making the associated perspectives easy for most readers to understand and digest.

Although this book primarily focuses on the textile goods we consume, I will also draw parallels with other industries to discuss the effectiveness of global sustainability efforts. For instance, global societies are investing trillions of dollars in specific areas such as the transition from the Internal Combustion Engine (ICE) vehicles to the Electric Vehicles (EV). Some readers might think that vehicles are unrelated to textiles and view this subject as out of context. However, they may be surprised to learn how closely the two are interconnected, as components like seatbelts, airbags, and seats are all textile materials. Moreover, the environmental impacts of any industry do not exist in isolation. They all influence our overall sustainability in one way or another.

Misinformation Sources and Causes

Another crucial perspective of this book is to highlight extensive misinformation prevalent on the internet. With recent advancements in the information technology, we increasingly rely on the internet for information. Google has largely supplanted encyclopedias and textbooks for a significant portion of the population. The advent of the ChatGPT and other AI-based platforms has further enhanced an easy access to both accurate information and misinformation.

Although I am not a software engineer and lack detailed knowledge of search engine mechanisms, my decades of experience using these tools have shown that many widely used soft tools can contribute to the spread of misinformation. For instance, search engine results appear to often prioritize popularity over accuracy - This phenomenon will be discussed in greater depth in **Chapter V: Solutions**.

While, in some cases, popularity may align with accurate information, this is often not the case in the realm of sustainability. Misinformation frequently overshadows true scientific data and accurate information on the internet.

One reason for this phenomenon, in my view, is the proliferation of self-proclaimed experts and organizations, operated or influenced by the misinformation generated by themselves. Areas related to the environment, sustainability, ESG, and climate change have become fertile grounds for individuals and organizations to claim their expertise and label themselves as experts without substantial qualifications. Consequently, there has been a myriad of unsubstantiated assertions, even from highly influential authorities – many examples will be presented throughout the book.

The problems created by these actions are detrimental to global sustainability efforts and must be addressed with the utmost priority. The widespread misinformation and erroneous claims impede genuine progress and undermine the integrity of the global sustainability efforts on the environment.

Structure of the book

I decided to take a slightly unconventional approach in organising the contents of this book. Given its focus on presenting science-based information about various sustainability claims and environmental initiatives, a more traditional structure would begin with basic educational contents before moving on to more practical subjects. Initially, I wrote the book in the order of: 1. Basic science of textile materials and manufacturing, 2. Lifecycle assessment, 3. Consumer-related subjects, 4. Misinformation crisis, and finally, 5. Solutions.

However, I chose to reorder the chapters, placing the consumer-related subjects first. This decision was made with the consideration that it might be easier for most readers without related knowledge or educational/industry experience to digest and follow through the contents. I believe this approach will help readers maintain a higher level of engagement by relating their personal experiences to increasingly scientific information as they progress. To support this objective, I included various *internal* references to related subjects, helping readers connect the dots with relevant scientific concepts in the dedicated discussions of the subject matters throughout the book.

Reference materials

To ensure the **credibility and transparency** of the information presented, this book includes a wide array of external references to **peer-reviewed scientific research** and validated sources. Rather than confining citations to appendices or endnotes, as is often customary in academic works, I have chosen a more reader-friendly method of citation.

Each reference is **embedded within the relevant section**, accompanied by **clickable links for e-book users** and **scannable QR codes for print readers**. This enables readers to verify information in real time, deepening both trust and understanding.

For those seeking more in-depth knowledge on textiles, particularly in the context of **thermal and winter applications**, which are among the most environmentally impactful due to elevated material and chemical demands, I encourage exploration of my companion book:

COLD WINTER WARM WINTER: Same Weather Different Protection – A Smart Guide to Staying Warm from Everyday Life to Advanced Textile Innovations, available globally via Amazon, including the following link and QR code:

https://www.amazon.com/COLD-WINTER-WARM-PROTECTION-INNOVATIONS/dp/B0FJY5464M/ref=tmm_pap_swatch_0?_encoding=UTF8&dib_tag=se&dib=eyJ2IjoiMSJ9.OIliAeUlbMDZ8WTV9oq_fQ.JvAvlMTMFPw0clYUzl1T0P2lPHXUBYmRXeCcQ84wdXw&qid=1753916438&sr=8-1

Industry Specifics

This book also includes **advanced technical discussions** intended for professionals in the textile and apparel industries, as well as for students pursuing specialized

13

studies in related disciplines. These sections are clearly labeled ***Industry Specifics*** to allow general readers to navigate the book without losing the continuity of the main content.

However, I acknowledge the inherent difficulty in neatly dividing material between "general" and "industry" audiences. The complexity of today's environmental challenges demands that **all stakeholders - consumers, students, and professionals - develop a stronger understanding of key issues**. In this spirit, I have made a **conscious effort to minimize such segmentation**, integrating technical depth in ways that elevate the reader's knowledge regardless of background.

***

Before delving into the book, I must ask readers to understand that my expressions may be blunt and direct. I feel that this is necessary due to the urgency for the global society to address significant challenges we face. While many people trust that societal authorities, such as government organizations, affiliates, and some of the largest NGOs on environmental matters, guide us correctly, I have observed that even highly authoritative organizations often make erroneous claims and engage in greenwashing. Throughout this book, I will illustrate numerous examples of such occurrences with candid expressions.

Credits

This book was written with the assistance of several individuals. From the outset, I recognized the risk of being confined by my own educational and professional experiences. Considering that the primary goal is to convey critical messages about environmental management and sustainability to as wide an audience as possible, the language used in the book must match the eye levels of the audience.

To address this, I sought fresh perspectives from the individuals without prior knowledge or experience in the textile industry. Their ability to comprehend the contents serves as a valuable indicator that the broader public will also find it accessible and understandable.

With this goal in mind, I asked the following individuals to partake in the assigned roles:

Paula Cevallos Agreda is currently in the master's program in the Management – Sustainability of HEC, Montreal and conducted research on some contents of the book, including the experiment on the comparison of power consumption between cotton and polyester in home laundry.

Anna Chumbe is MA student in the Comparative Literature program of University of Montreal. Anna worked on the proof reading in English and helped in editing.

Hyeonji Yu is currently a student at Chungbuk National University, South Korea and worked on some graphics presented in the book.

I sincerely thank the intelligent individuals who helped me on this important journey. Without your help, this book would not be what it is.

Now, let us begin our journey to becoming better informed and educated.

Chapter I

*

Consumers, Be Aware and Beware!

Pre-Chapter commentary

Quote of the chapter

"Given the prevalence of greenwashing, even by some of the most trusted organizations globally, the public's overly simplistic approach to various sustainability claims does little to benefit the environment. It is crucial for consumers to take a more critical stance and demand transparent, comprehensive information in order to make informed decisions that truly benefit the environment."

***

Understanding the environmental implications of various textile materials, such as cotton and polyester, begins with examining their origins, whether grown in cotton fields or synthesized as synthetic fibers. By reviewing manufacturing processes to becoming the finished products we purchase and tracing their journey during and after consumption, we gain a critical knowledge base for evaluating the true environmental impacts of different material types.

This chapter aims to provide the essential information required for such an analysis, offering readers valuable insights that may influence not only on the sustainability aspects of their mindsets, but also in other areas of their lives, such as making purchasing decisions and deciding on the maintenance practices of their textile goods.

While some scientific details presented in this book may appear challenging, readers should not feel overwhelmed as I made extra efforts to explain the related topics in easily digestible manners so that even those who do not have backgrounds in science or engineering can grasp the concepts effectively. By completing this book, you will acquire a comprehensive understanding of textile sustainability.

Life from Beginning to End

Textile goods are indispensable components of human life, encompassing all stages from the birth to the death. This intrinsic connection to humanity underscores the textile industry as one of the oldest in existence. A compelling anecdote is the biblical account of Eve, who, feeling ashamed of her nudity after consuming the forbidden fruit, sought to cover herself with a leaf from a tree, symbolically marking the inception of the textile industry.

Humans, in ancient times, used various materials such as animal fur, leather, leaves, etc., to cover their bodies. From an evolutionary perspective, early humans may have utilized natural materials for warmth and protection after progressively losing their body hair during the evolution.

In many cultures, the significance of textiles extends beyond the life on earth. For instance, it is a common practice for individuals to select the garments they wish to be buried in, reflecting their beliefs in a well-prepared afterlife. The elaborate burial practices of the ancient Egyptian Pharaohs, who had meticulously planned to have their bodies wrapped in multiple layers of linen during mummification and placed in the archaeological wonders of Pyramid, illustrate the fundamental desires of humans even after death. These historical and cultural perspectives highlight the longstanding and integral role of textile goods in human civilization, from the ancient times to the present.

Humans as Social Animals

Throughout our lives, from infancy to old age, we consume a vast array of textile goods in significant quantities. The consumption of textile goods is intricately linked to one of the fundamental human instincts associated with social behavior. Anecdotally, we can easily imagine that the biblical figure Eve selected the most visually appealing leaves to cover herself, continuously seeking aesthetically pleasing materials to enhance her appearance for Adam while the later reciprocating such effort. Textile goods have long served as essential tools for self-expressions and social interactions, reflecting our nature as social beings.

Material Consumption Patterns

Most individuals possess a considerable degree of experience and knowledge regarding which types of fibers align best with their lifestyles and values. For instance, I personally prefer cotton for socks and wool for business suits, while I favor synthetic fibers for nearly all other uses. The reasons behind my preferences and purchasing decisions form a significant part of this book, as they are based on comprehensive scientific analysis not only on the part of performance metrics but also on the environmental sustainability. I understand that a large proportion of consumers would prefer to purchase high fashion goods made with natural materials, if financially feasible. This preference may be partly due to a perceived quality and comfort, and partly on environmental concerns over synthetic materials.

To provide a context to this discussion, I present the pie chart of the worldwide fiber consumption in *Fig.1*.

Fig.1 Worldwide Fiber Consumptions 2020

Fiber production share

https://www.textiletoday.com.bd/global-polyester-fiber-market-share-and-scopes-for-bangladesh

Dominance of Polyester

Polyester dominates, accounting for 52% of the total fiber consumption. Cotton follows with 24.2%, while all other fibers constitute relatively minor portions. Notably, my personal wardrobe would likely reflect a higher percentage of polyester than the global average.

Some readers may wonder why polyester is so prevalent, especially when many people express a preference for textiles made from natural materials. Various reasons behind this disparity are explored throughout this book, with discussions encompassing the entire lifecycles of key

20

material types from farming to manufacturing, consumption, and disposal.

Given the limited space available, a detailed analysis on every fiber type is not possible in this book. Instead, an emphasis is placed on the two dominant fibers: cotton, representing natural fibers, and polyester, representing synthetic fibers. Together, these two account for close to 90% of the global fiber consumption.

Historical Analysis

In the meantime, historical data analysis reveals that the fiber compositions do not change drastically year over year. A supplementary chart from 2018, with a slightly different categorization of fibers, provides similar splits.

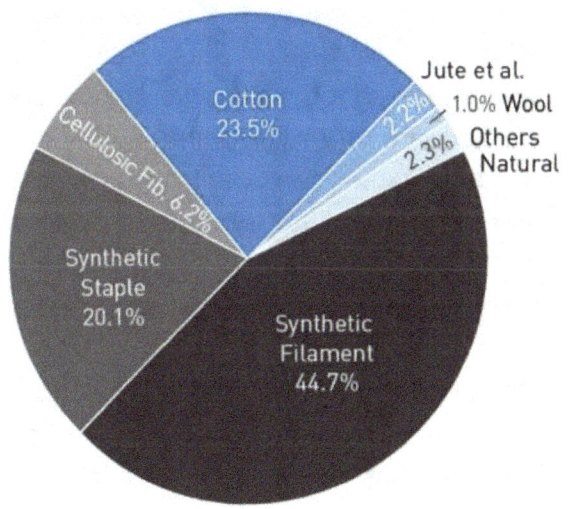

Fig.2 Worldwide Fiber Consumptions 2018

https://textile-network.com/en/Technical-Textiles/Fasern-Garne/Forecast-world-fibre-production

For example, in **Fig.1**, cotton constituted 24.2% of the total fiber consumption while it is 23.5% in **Fig.2**. Given that the difference is only 0.7% between 2018 and 2020 and there is an inherent margin of error in these types of statistic figures, this variation is considered negligible. Conversely, the combined share of the synthetic fibers in the 2020 pie chart (**Fig.1**) was 62.2%, comprising polyester (52%) and all other synthetic fibers (10.2%). In 2018, the synthetic fibers made up 64.8% of the market, categorized in the synthetic filament (44.7%) and the staple (20.1%). This small difference of 2.6% on the synthetic fiber consumption can also be viewed as negligible and perhaps within statistical margin of error, indicating that the consumption of the synthetic fibers has also remained relatively stable year over year.

Consumption Rate and Economic Development

The overall quantity of textile consumption is closely tied to economic and technological development. Most textile goods have become more affordable and accessible than ever before, even when considering inflation and shifts in monetary values. This trend is particularly evident for the textile products produced through industrialized manufacturing systems. For instance, the cost of producing a common cotton shirt has decreased significantly over time, thanks to numerous advancements in farming techniques, manufacturing efficiency, and distribution methods, particularly since the industrial revolution.

The textile consumption also correlates strongly with the income levels of consumers. In underdeveloped and developing regions, as the income rises, the consumption tends to increase proportionally. However, in highly developed countries, once the income reaches a certain threshold, the rates generally stabilize.

Essential Needs for Education

Given the critical role textile goods play in our daily lives including the environmental sustainability, it is vital for everyone to gain a basic understanding of key aspects related to these materials. While some readers may already be familiar with the lifecycles of the textiles they consume from raw materials to manufacturing, distribution, consumption, and disposal as well as performance metrics and environmental impacts, many others may lack this knowledge.

With my educational and professional background, I am frequently engaged in casual discussions on a wide range of textile-related topics, from environmental impacts to quality and maintenance, with those around me. From these conversations, I often encounter people who feel confident in their understanding of various textiles and their ability to make informed purchasing decisions. However, in my view, significant gaps are often found between their perceived knowledge and scientifically sound and sustainable practices. For example, an intentional choice of a supposedly more sustainable material can sometimes lead to unintended environmental harms. This chapter invites readers to delve into these issues and reassess whether their confidence aligns with the practices that truly benefit the environment.

***

Unfortunately, many textile companies and brands take advantage of consumers by promoting certain products with notion-based claims or pseudoscientific features that fail to deliver in real-world conditions. Similarly, a significant number of environmental advocacy groups spread misconceptions due to a limited understanding of underlying science. This can result from a lack of thorough research or insufficient diligence in verifying the accuracy of the information they disseminate.

The term "greenwashing" is often used to describe these undesirable practices, particularly when it comes to environmental matters. As a result, both the public and consumers are often misguided into accepting misinformation as scientifically valid truth. In this chapter, various examples will be explored to critically examine how consumers are unknowingly misled by numerous cases of corporate misconduct and ill-informed advocacy.

Section 1 Unfortunate Consumers

The Section 1 explores several critical topics of misinformation and their sources, with a particular focus on the consumer perspectives. Initially, I will present broader viewpoints on global environmental management, highlighting how various environmental interest groups and activists often misplace blame on consumers and textile companies based on unscientific or invalid notions. Afterward, the discussion will shift to legitimate concerns about poor business practices within the textile industry and the consumer market, including those involving some of the largest and most reputable companies worldwide.

While I made an effort to provide detailed explanations backed by scientific information in relevant sections throughout the book, some of the brief descriptions and statements in the following discussion topics in the beginning may challenge many readers' existing beliefs. However, I encourage the readers to approach these topics with an open and inquisitive mindset, as there will be ample opportunities for deeper discussions and the scientific analyses of important environmental issues in various subjects ahead.

Consumers Targeted

In recent years, consumers, especially those in the Western world, have been the focus of many environmental interest groups, who criticize the overconsumption of apparel, particularly under the context of so-called "fast fashion". However, various research indicates that the global growth in textile consumption is primarily driven by *"the global population and income growth in emerging economies, with very little contribution from volume growth in Western markets"*

https://emf.thirdlight.com/file/24/uiwtaHvud8YI G_uiSTauTUH74/A%20New%20Textiles%20Econ omy%3A%20Redesigning%20fashion%E2%80%9 9s%20future.pdf

Major Development in Consumption Trend

As mentioned earlier, the consumption rates of textile goods are closely related to basic human instincts. Throughout history, people have always sought to look good and express their identities through their appearances. As the public gained more economic power, it was a natural occurrence that the consumption rates increased. However, for the first time in human history, we are witnessing a plateau in the consumption rates in some affluent countries, as the satisfaction of personal desires begins to align with the quantity of goods people consume.

Furthermore, there have been encouraging signs that growing environmental awareness among the public is contributing to this significant shift in the consumption patterns. As this awareness grows, there has been a movement urging citizens in affluent countries to reduce their consumption of textile goods, while the global community accommodates a natural consumption growth in underdeveloped and developing nations.

Need for Effective Efforts

Although the trends in some developed countries show positive signs, significant challenges remain with textiles. The cumulative effects of human living over thousands of years, particularly accelerated by the rapid industrialization and population growth, pose an imminent threat. This urgency is further underscored by the toxic impacts most textile goods create throughout their entire lifecycles.

Effectively addressing the environmental impacts of textiles to achieve an overall reduction is critical. Simply blaming global consumers at large for overconsumption is unlikely to be effective. Instead, more successful strategies will involve well-designed approaches that target consumers in different economic zones, combining education and methods that appeal to their psychological consumption behaviors and basic instincts. These tailored strategies can help create more positive environmental outcomes. This approach will be further explored in **Chapter V: Solutions**.

Companies Targeted

Many environmental advocacy groups often hold various companies in the global textile industry responsible for the environmental and sustainability challenges we face today. While it is undeniable that these companies create direct impacts on the environment through their operations, by fixating on corporate responsibilities, often with scientifically invalid bases, we risk overlooking deeper, instinct-driven consumer behaviors that contribute more majorly to these issues. After all, consumers ultimately make purchasing decisions based on their own free will.

Virtue of Profit-Seeking Organizations

From the perspective of the profit-seeking organizations, it has long been viewed as a virtue to maximize their business outcomes in the form of profits. When there is a consumer demand, there is always a seller willing to create a system to meet those needs and turn it into profits. This is how the economics has functioned for thousands of years. Operational strategies of these companies might include a wide range of factors like point of sale, shipping method, marketing, technological development, and more, many of which are now under environmental scrutiny.

Ill-Conceived Targets

These targeting acts, however, often lack proper context. For example, blaming a "fast fashion" company for using air cargo, which generates higher carbon emissions than sea freight, must be contextualized against the environmental impacts of "slow fashion" companies. The later form of business often operates large storefronts that consume substantial energy for lighting, heating, and more, along with the carbon emissions generated by their employees commuting to and from these locations and the resources needed to maintain such corporate systems. In reality, we lack sufficient data to fairly compare the overall environmental impacts of these different operational models. This example illustrates how many environmental advocacy groups frequently present information selectively, focusing on the points that support their arguments without considering the full picture.

Hidden Sides

In the public eye, some companies may seem more ethical by adopting environmentally or socially responsible business practices in certain aspects of their operations. However, throughout my career, I have often seen that even some of the most reputable global companies prioritize promoting their positive images, while failing to live up to their claims, promises and often engaging in what can only be described as greenwashing. In various discussion subjects of this chapter, I will present several concrete examples of such corporate misconducts.

Fair and Comprehensive Judgement

Conversely, conventional judgments on various environmental and sustainability issues often rely more on consumers' general perception of a certain brand and/or the retail values of their goods (expensive vs. inexpensive), business practices, etc., rather than on true environmental impacts based on accurate science. In more recent years, certain trade conflicts have led to influencing public judgments on the goods originating from some countries under the context of the environment and sustainability.

However, it is important to emphasize that the true environmental impacts are rarely confined to singular or overly simplistic aspects and notions. Our evaluations must be comprehensive and grounded in accurate scientific analysis. Without a balanced and fair approach to the judgments, targeting specific entities or practices under the banner of sustainability can be misguided and may not lead to desired improvements in environmental outcomes.

Overproduction Issues and Safety Hazards

For example, recent overproduction issues in China have become a major topic in the business world and are often connected to various environmental concerns. Furthermore, there have been numerous reports of safety hazards associated with products purchased from online platforms that offer extremely low prices. These products, which include toys, personal accessories, plastics, textiles, cosmetics and more, often contain harmful chemicals that exceed the safety limits set by different countries.

This issue highlights a significant lack of quality control and moral ethics among the manufacturers, exporters, importers, and platform operators involved in such incidents. These safety hazards not only pose risks to consumers' health but also have detrimental environmental impacts.

Regardless of the environmental concerns, however, such business practices should not occur in the first place as they are illegal and intolerable. Unfortunately, these illegal practices are not limited to low-cost value chains as they have also been found, though presumably less frequently, in products sold by even reputable companies with higher price tags on their goods. These incidents may happen with or without the knowledge of some or all of the involved parties, primarily driven by profit motives and/or lack of efforts.

Need for Accurate Assessment

In my view, trade conflicts, unfair business practices, extremely low prices and safety hazards should remain focused on economic, legal, and political debates, rather than being conflated with environmental and sustainability issues. For example, if a country determines that certain goods sold on a particular platform are likely to contain harmful chemicals that threaten the health and safety of its citizens, necessary actions should be taken to prevent such risks, independent of other considerations. While it is true that environmental topics tend to capture public attention more effectively and elicit stronger reactions these days, this approach can obscure the core issues that need to be addressed for the environmental sustainability.

1.1 Material Choices and Care Methods

One of the biggest areas of misunderstanding in textiles, driven by widespread misinformation, involves the public perceptions of different material types and their environmental impacts. For example, many people believe that natural fibers like cotton, wool, and silk are more sustainable than synthetic fibers like polyester, Nylon, and polypropylene. This perception is rooted in the following three common perspectives:

A. **Origin Factor**: *"Synthetic fibers are produced from petroleum sources, creating carbon footprints, whereas natural fibers are sourced from the nature"*.

B. **Decomposition Factor**: *"Synthetic fibers persist in the environment much longer (some argue indefinitely), while natural fibers decompose quickly"*.

C. **Microplastic Factor:** *"Synthetic fibers shed microplastics, contaminating our food sources, leading to bioaccumulation in humans and animals, and causing health issues"*.

These seemingly straightforward arguments resonate with the public and have shaped the global environmental mindsets and associated efforts. While some of these points are accurate in isolation, they often lack numerous critical contexts needed for a more comprehensive assessment of the true environmental impacts between the two fiber categories. Below are some of the examples:

A.1 **Origin Factor**: *"Synthetic fibers are produced from petroleum sources, creating carbon footprints, whereas natural fibers are sourced from nature."* : This argument critically overlooks the significant energy consumption involved in farming natural fibers and, in some cases, the amount of the carbon footprint created in natural fibers can be multiple times higher than that of synthetic fibers - Detailed discussions on carbon emission comparisons amongst different fiber types will be presented with a set of scientific data from a research paper in **Chapter III: 4.4 Carbon Footprint Index (CEI) Analysis.**

B.1 **Decomposition Factor**: *"Synthetic fibers persist in the environment much longer (some argue indefinitely), while natural fibers decompose quickly."*: The notion that natural fibers decompose quickly considers the breakdown of the natural elements like the cellulose of cotton and the protein of wool. However, it fails to account for the residual chemicals in textile waste. The quantity of harmful chemicals contained in the fibers, their toxicity and persistence in nature are mostly overlooked, yet critical factors for evaluating the overall environmental impacts. This topic is one of the central themes of this book and will be explored in depth throughout the book, including the dedicated **4.2 Decomposition Myth** discussion later in this chapter.

C.1 **Microplastic Factor:** *"Synthetic fibers shed microplastics, contaminating our food sources, leading to bioaccumulation in humans and animals, and causing health issues."*: While synthetic fibers do release microplastics throughout their lifecycles, this argument critically overlooks the issues from natural fiber dust. Natural fibers, like synthetics, shed dust during their lifecycles. The quantity of the natural fiber dust and the amount of harmful chemicals contained in it compared with the synthetic microplastics may surprise many readers - A detailed analysis on this subject will be provided in **3.2 Dust of Natural Fibers** later in this chapter.

Furthermore, many other critical factors are missing from the global perception of the environmental impacts between natural and synthetic fibers. To keep the discussions more relevant for each subject throughout the book, I will present these factors in the appropriate sections and subsections. In this section, I focus on the topics more closely related to the consumption of textile goods, starting with the discussion on the care methods:

1.1.1 Care Method Overview

Care instructions on the woven labels attached to most textile goods vary based on several factors, with the most significant being the recommended cleaning method, either home washing or drycleaning, often determined by material types, whether natural or synthetic. These methods are intended to help consumers preserve the appearances, shapes and fits of their textile goods in their original conditions for as long as possible. Additionally, there are other factors related to the care and maintenance that are not typically incorporated in the care instructions. This discussion will explore these important considerations, which not only affect the longevity of textile goods during use but also create significant environmental impacts.

Micro-Orgasmic Activities and Frequencies of Care

Material types can heavily influence the care frequencies during consumption period. For instance, cotton shirts and rags tend to develop odors more easily than their polyester counterparts, a phenomenon familiar to many. This observation aligns with the scientific understanding of the natural fibers and their chemical compositions, such as the presence of hydroxyl groups in the plant-based fibers (cotton, linen, etc.) and proteins in the animal-based fibers (wool, silk, etc.), which inherently offer high chemical reactivities with odor-causing microorganisms - A more detailed exploration on this topic is provided in **Chapter II: Basics**. It is therefore a common practice to wash or dry-clean textiles made with natural materials more frequently than their synthetic counterparts.

Color Fading and Physical Deformation

The natural fiber textiles are often recommended for drycleaning due to their higher susceptibility to color fading and physical deformations, such as shrinkage and stretching, when washed with water at home. These issues stem from the higher chemical reactivity of the natural fibers, combined with their lower inherent strength. This is particularly true for the animal-based materials like wool, silk, leather and down, which frequently come with drycleaning instructions - The toxic nature of drycleaning chemicals and their environmental impact will be discussed in detail later in this section.

***

In the following discussions, we will explore the principles behind different care methods for various fiber types and their corresponding environmental effects.

1.1.2 Natural Materials and Care Methods

Numerous natural materials are used in textiles, primarily categorized as either animal- or plant-based. In this subsection, we will explore three of the most commonly used natural materials: Cotton, Wool and Down.

A. Cotton

Cotton's affinity for water and the heightened chemical reactivity of its hydroxyl groups present significantly different characteristics compared with synthetic fibers in maintaining the appearance, shapes and fits of the textiles made with the material. For instance, most readers may have noticed that their cotton shirts tend to lose color more readily than their synthetic counterparts after multiple home washes. In knitted products such as sweaters, T-shirts, etc., it is also common to observe necklines becoming loose especially if hung-dried or a considerable shrinkage occurring when the temperature setting of washing machine and dryer was too high. All these phenomena can be explained by the chemical compositions and strength factor of cotton in comparison with its synthetic counterparts.

While most textiles containing cotton can be maintained with careful home wash, some individuals opt to dry-clean their valuable cotton garments to prevent color fading and deformation mentioned earlier. Since drycleaning does not involve water, it significantly reduces the risks of such occurrences.

B. Wool

Wool fibers offer a unique surface structure with the scales in the linear alignments of the fiber's length direction, as shown in *Fig.3*.

Fig.3 Scales on Surface of Wool Fibers (coarse and fine)

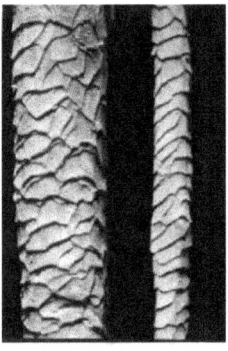

https://www.silhouettetailoringstudio.com/shrinkage/

Similar to fish scales, the scales on wool protect the internal protein structure beneath. Unlike the protein of fish, however, wool's protein molecules absorb a large amount of water and swell when submerged in water. As a result, the outer edges of the scales extend and protrude more outward. During a home wash with water, the physical agitation inside the washer causes the scales, aligned in opposite directions, to interlock with each other as they act like tiny hooks. This process, known as "felting", leads to shrinkage of woolen items in home laundry. Once the felting occurs, it cannot be undone, effectively ruining the item. This is why the care instructions of woolen goods typically recommend drycleaning.

In some cases, woolen items are labeled as washable in water. However, this requires a process that dissolves the tip parts of the scales in strong chemical solutions. While it can eliminate the need for drycleaning, it introduces potentially much more harmful chemicals during manufacturing process.

C. Down

The down-feather material, commonly known as "down" and originated from various types of birds, presents a higher level of complexity than other textile materials, primarily because it is typically used in more intricate constructions that involve multiple components, such as enveloping outer fabrics and insulating layers which contain the material. These complex structures can be easily understood when comparing between a winter jacket with down material located deep inside the structure, thus invisible, and a single-layered shirt or pants.

Instructional Variations: Textile goods that utilize down materials as insulator fall into two categories based on their care instructions - those that are instructed to be drycleaned and others recommended for home washing. This discrepancy often leads to questions and confusion, particularly given that the primary goal in both scenarios is to clean and maintain the performance of the valuable down material.

Dry Cleaning Recommendations

Following factors are considered for the drycleaning recommendations:

Drying Time and Fabric Treatment: Washing textile goods containing down materials at home can require multiple drying cycles unlike most other items which will dry in just one cycle. While the complex structure mentioned earlier is a contributing factor, this extended drying time can be primarily due to the plasticized fabrics used to prevent down leakage - The structure of woven fabrics enveloping down material creates small holes at the intersections of the crossing yarns, known as warp and weft. Although these holes are small in sizes, they are larger than the fine hairs of down feathers and can become the channels of escape, as illustrated in *Fig.4*.

Fig.4 Down Leakage

To mitigate this, the fabrics used to create the containment of down materials, commonly known as "down bags", are often plasticized to close off the holes as shown in *Fig.5*.

Fig.5 Plasticization and Down Leakage Prevention

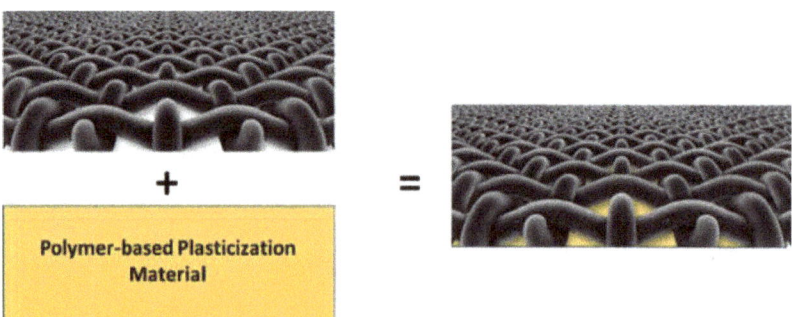

Polymer-based Plasticization Material

Despite this effort, consumers still experience down leakage often through the needle holes as depicted in **Fig.6**.

Fig.6 Down Leakage through Needle Holes

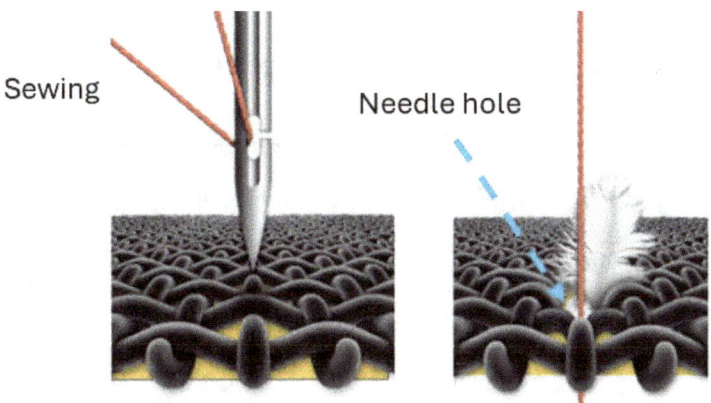

This plasticization covers much of the fabric, significantly slowing moisture evaporation rate during drying process by blocking the escape of water molecules, illustrated in **Fig.7**.

Fig.7 Water Molecules Bouncing Back

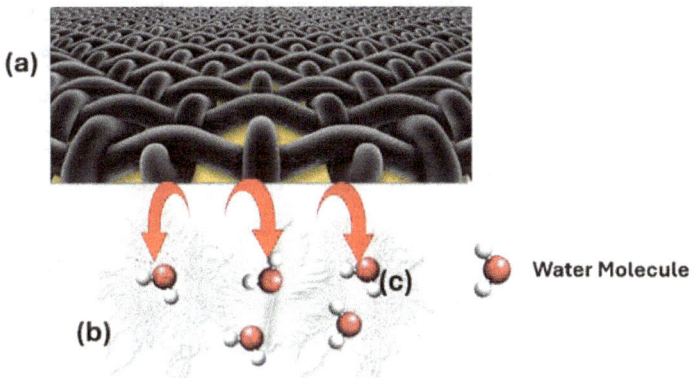

(a) Plasticized fabric surface (b) Down material
(c) Water molecules resisting evaporation

Consequently, drying a down-filled product at home can be a half-day chore, requiring multiple drying cycles, often with tennis balls to agitate the down material, ensuring that it dries thoroughly from the inside out. Inadequate drying can leave the material damp deep inside the structure, creating an ideal environment for microbial growth, fueled by the protein in the down feather and the nutrients from our own sweat, such as urea, ammonia, and fatty acids. Although many people may not recognize, "down" items such as garments, blankets, sleeping bags, etc., can be the sources of various health issues if improperly maintained. Acknowledging these challenges, many brands recommend drycleaning.

Appearance: Drycleaning instructions can be motivated by other reasons. For instance, if a down jacket contains metallic components such as badges, zippers, buttons, and snaps, washing it with water and detergents can lead to significant discoloration and unwanted aesthetic changes. Drycleaning helps mitigate this issue. In such cases, the drycleaning instruction is unrelated to the care or performance of the down material. However, some notable brands avoid using metallic components altogether. Instead, they use embroidered textile-based badges, along with plastic components for zippers, buttons, snaps, etc. This allows them to safely recommend home washing to their consumers, reducing the significant environmental impact of toxic drycleaning chemicals while also offering considerable cost savings for maintenance.

Warmth Performance affected by Cleaning Methods: Down's delicate structure, made up of numerous fine hairs, is more vulnerable to damage from the physical agitation of home washing mechanisms. When down feathers absorb water, they become considerably heavier, increasing physical forces acting on them during washing, which raises the risk of damage and ultimately reduces their warmth performance. Drycleaning, which does not involve water, is gentler on down. However, the effects of infrequent home washing, such as once or twice per winter season, often go unnoticed by untrained consumers, as a sufficient amount of down is typically inserted during manufacturing to maintain a certain level of warmth over a considerable period of use.

Effectiveness of Cleaning Methods: In the perspective of the effectiveness of the two different cleaning methods, many pieces of conflicting information are found on the internet, making different claims, some insisting that down materials need to be drycleaned while others favor home wash. Each argument proposes a reasoning or two. Rarely, however, these discussions include the nature of contamination in relation with the care methods.

In essence, home wash is effective if a contaminant is soluble in water. If the contaminant is not water-soluble, certain drycleaning chemicals may be more effective at removing it.

Nature of Contaminants: A large proportion of contaminants for down materials are water-soluble, making home washing with suitable detergents sufficient for their removal. However, exceptions exist. For example, cooking oil splashes onto clothing. Depending on the type of oil and the nature of the contaminated fabric, these stains can be stubborn and resistant to home washing. Dry cleaners may use specific chemicals that are effective for such oil-based contaminants.

That said, in everyday use, it is unlikely that down materials will be exposed to water-insoluble contaminants, as they are typically protected by layers of enveloping fabrics. Even in an unlikely scenario where someone cooks while wearing a winter jacket and oil splashes onto it, the chances of the oil penetrating through the outer layers and contaminating the down are minimal. The most common contaminants for the down materials in such protective structures are the compounds contained in our sweat as mentioned earlier, as they penetrate through the fabrics and reach the material. These substances are water-soluble and can be effectively removed through home washing.

* Special mention – Leather, while it is one of the most commonly used natural materials, is not covered more extensively in this discussion largely because they are invariably instructed for drycleaning. It is a common knowledge that home-washing leather goods can create significant discoloration. This phenomenon relates to the inherent chemical reactivity of both the leather itself and the chemicals used to colorize it, which will be further discussed throughout the book.

Chemical Toxicity of Drycleaning Chemicals

A surprisingly large number of people are unaware of the toxicity associated with drycleaning chemicals. Perchloroethylene (PERC), one of the most commonly used solvents in drycleaning around the world, is a reproductive toxicant, neurotoxicant, potential human carcinogen, and a persistent environmental pollutant. I vividly recall one of my textile engineering professors warning our class, "*If you live in a building where there is a drycleaning shop, move before it kills you!*". His opinion, while it may be debatable given many variables from one building to another with different exhaust systems, etc., was to alert the students of the nature of most solvent-based cleaning chemicals as odorless gas that spreads quickly and widely, making it particularly hazardous.

Nature of Solvents: Although efforts have been made to develop less toxic solvents for drycleaning, ongoing debates persist about the sustainability and safety of these alternatives. Drycleaning solvents, by their very nature, are chemically potent enough to alter the states of contaminants, detaching them from contacting textile materials by chemical interactions without water. This potency inherently makes them hazardous to both humans and the environment.

1.1.3 Synthetic Materials and Care Methods

Different synthetic materials exhibit varying characteristics in terms of their environmental impacts. While detailed information on these overall environmental effects will be thoroughly examined in **Chapter II: Basics**, one significant advantage of most synthetic materials is that they typically do not require drycleaning. This distinction, compared to their natural counterparts, stems from a significantly lower chemical reactivity, which gives them a strong resistance to color fading. Furthermore, they offer much higher strength, enabling them to withstand physical forces encountered in home laundering with water.

I own two polyester T-shirts in black that I purchased nearly twenty years ago. They are among the most frequently worn and home-washed items in my wardrobe. As the black is a color that shows fading easily and clearly, I compare these polyester shirts with other items made with cotton over the years and observe the progression of color fading – Readers may make similar observations with their own garments. The polyester shirts have maintained their color and shape exceptionally well, while the cotton items have noticeably faded over much shorter periods of use.

***

After reviewing several key factors associated with different care methods and their environmental impacts across various textile material types, following discussions will focus on specific technologies used in consumable textile products and their potential environmental impacts:

1.2 Water Resistance & Breathable Technology

Many outdoor enthusiasts may recognize this technology, as it is widely used in outdoor apparel and is also commonly found in workwear, tactical gear, and other applications. This subsection explores how the technology functions and its potential environmental impacts.

Definition: Before diving deeper into this topic, it is essential to clarify a common misunderstanding regarding the terminology. "Waterproof" is a loosely applied term that relates to the performance of the fabrics with this technology. In a more precise scientific description, it refers to a resistance up to a certain level of water pressure. Therefore, the term "water resistance" is more accurate than "waterproof", as beyond a certain pressure threshold, the fabric will leak water. Applying a similar logic, many other fabrics can be labeled as "waterproof". For example, tightly woven synthetic fabrics without this technology can resist water up to a certain pressure. The key difference lies in how much water pressure a fabric can withstand before allowing water to seep through. Despite the discrepancy, the term "waterproof" for the fabrics employing this technology has been used in the industry for decades and is widely accepted by both sellers and consumers.

"Breathability" refers to the fabric's ability to allow moisture-laden air from the body to pass through (while providing a high level of water resistance in the context of this technology platform), providing dry comfort to wearers.

Confusion with Water Repellency: Additionally, some consumers confuse water resistance with "water repellency". Companies promoting "waterproof" feature often use images showing water droplets beading on the surface of a fabric, the phenomenon unrelated with the "waterproof" claim. While both water repellency and waterproof features give the appearance of resisting water, they differ significantly in the chemicals used and the areas where the chemicals are applied – The "water repellency" will be explained after the ongoing discussion.

Technology Principles: The "water resistant and breathable" technology involves applying a polymeric membrane, typically made of the expanded-polytetrafluoroethylene (ePTFE), to the fabric's backside (the side opposite to the weather-facing surface). This membrane contains countless tiny pores that are small enough to prevent water droplets, such as those from rain, held together due to the surface tension of the comprising water molecules, from passing through. However, the pores are large enough to allow moisture vapor from sweat, which consists of smaller water molecules, to escape, achieving both the water resistance and breathability.

That said, under enough pressure and depending on the size of the pores, the water on the fabric surface can eventually overcome its own surface tension and penetrate through the membrane, allowing it to reach the wearer's skin. This phenomenon is better understood by reviewing a typical test method used to evaluate the performance metrics.

Test Method: A piece of fabric is placed flat within a cylinder that separates the top and bottom compartments with the fabric dividing the two. Water is then poured into the top compartment, while the bottom compartment remains empty to allow the observation of water droplets passing through the fabric. Test equipment typically includes a pressurization system that simulates water pressure at specific levels, such as 5,000mm (or 5 meters / approx., 17ft), 10,000mm (or 10 meters / approx., 33ft), or higher.

As the water pressure increases, the fabric eventually allows the water to seep through, and droplets begin to fall into the bottom compartment. The tester records the water pressure at the point when a certain number of droplets are observed, which determines the fabric's water resistance level. For example, a 10,000mm rating indicates the fabric can withstand the equivalent of 10,000mm of water pressure.

Effectiveness in Extreme Conditions: This technology has gained a widespread popularity globally, with many people choosing garments and outdoor gear that feature it. For avid outdoor enthusiasts or professional explorers facing extreme wilderness conditions, high levels of water resistance can be crucial for survival. For instance, a 10,000mm performance level can be lifesaving in places like the Himalayas or the South Pole.

Phenomenon in South Korea: In South Korea, a particular brand known for this technology has become so popular that it is now applied in a wide range of apparel items, including large quantities of everyday clothing, regardless of wearers' activity levels or weather conditions they typically encounter. Students, grocery shoppers, casual walkers, and light hikers, including my own friends and family, frequently wear garments with this technology. During a business trip to South Korea, I visited a department store with my family and my sister was looking for a new jacket with this water-resistant feature, more specifically, one with the famous brand's logo. When I asked if she understood the technology and why she needed it, her response was simple: *"It is widely popular and everyone around me wears it. There must be a good reason for it, and it should apply to me."* Given that she is a working mom whose outdoor activities rarely extend beyond grocery shopping, it was clear to me that her desire for the feature defeated the purpose of this technology.

Phenomenon in Canada: More recently, I had a similar experience with a friend in Canada, an avid runner who regularly participates in half and full marathons. One day, he joined in a light hiking event I organized on a sunny day, and was wearing a pair of running shoes that incorporated the same brand as the one my sister was looking for. The brand's name, prominently displayed on the sides of the shoes, caught my attention. When I asked about his understanding of the technology, his response was remarkably similar to my sister's.

I have noticed this trend in many other countries I travel to, especially in affluent regions, where this expensive technology is often used in everyday clothing.

Practicality of Performance: One internet blog I came across claimed that a 10,000mm water resistance rating is equivalent to standing under a heavy waterfall for 30 minutes. Although I could not verify the accuracy of this claim, given such vague term as "heavy" waterfall, it is easy to understand that 10,000mm (10 meters or 33ft) is an extremely high level of water pressure for most people in their everyday activities. Personally, as a middle-aged man who considers his lifestyle to be more or less average in the general population, I have never encountered a situation where such a high level of water-resistant performance was necessary, even including during my younger years when I frequently enjoyed outdoor activities like intensive hiking in mountains and camping in Korea.

Sweat Challenges: Some of my friends and family members have mentioned that when they wear their jackets with this technology, they often feel a quick buildup of sweat inside, even during a short hike. This contradicts the high breathability perception consumers gain on these products as a result of associated promotional information.

Scientifically, this observation is accurate. When the level of breathability cannot keep up with the amount of sweat produced by the body, the vapor accumulates inside the garment. How quickly this happens depends on various factors like wearer's metabolism, activity level, health, and weather conditions. While the membrane contains numerous tiny pores, the rest of the fabric is blocked for the vapor's escape. If the sweat production exceeds the membrane's capacity to release it, the accumulation occurs.

Lack of Understanding: Choosing between increased water resistance with reduced breathability offered by this technology and maximizing breathability without this technology is a personal choice. However, in my opinion, preparing for extreme weather events combined with prolonged outdoor activity, that may never occur for many consumers like my sister, might not be the most logical approach. While this technology can be valuable in situations where highly pressurized water is a concern, such as in footwear, where our feet bear the entire body weight on a small surface area, or in the areas like elbows, knees, and buttocks of ski wear or workwear for specialized job functions involving kneeling, elbowing on wet surfaces, etc., its practicality for everyday clothing is what concerned me during my conversation with my sister at the department store. Unfortunately, many people are unaware of the trade-offs including the environmental impacts of the technology and are often influenced by social trends rather than their actual needs.

Environmental Impacts: Applying a polymeric membrane to the backside of fabrics, a process often known as "plasticizing", makes these fabrics unrecyclable with currently available recycling technologies. The fabric and membrane create a tight bond, which cannot be separated, significantly compromising the recyclability of these materials. Since successful recycling relies on the purity of the materials in question, this technology could result in large quantities of fabrics becoming unrecyclable, even though they could otherwise be efficiently processed using the **Clean Recycling Initiative™** technology platform. More detailed discussions on this platform will be provided in **Chapter V: Solutions**.

Additionally, materials like ePTFE used in this technology are incredibly durable. They are dense polymers and are some of the heaviest materials on Earth by volume. While I will address the widespread misunderstanding about plastic degradation later in this chapter, it is important to note that some plastics last longer in nature than others and their densities play a key role. As a result, textiles incorporating this technology may persist in the environment for much longer than the equivalents without it.

Debates: I have witnessed many debates regarding the environmental impacts of this technology. One stakeholder, in particular, made an argument that the technology extends the lifespan of the fabrics by making them stronger, assisted by the membrane, thus positively impacting the environment.

In analyzing these claims and debates, it is important to note that the fabric strength is rarely a reason contributing to the disposal of textile goods these days, particularly in affluent Western countries. This is evident when examining textile items deposited in community centers, recycling bins or donation depots as they are often in wearable conditions, sometimes needing only minor repairs like fixing loose buttons or broken seamlines, more of which will be discussed in the **Unfortunate Disposal** later in this chapter.

Consumer Awareness: When purchasing everyday textiles, my decisions revolve around the likelihood of encountering certain weather events and my plans for dealing with them. Based on these considerations, my personal shopping habits prioritize efficient sweat removal and the environmental impacts of the goods I purchase. As mentioned earlier, I have never encountered a situation where 10,000mm of water resistance was essential. In the event of sudden, heavy rain, I seek a shelter. For a forecasted weather event, I carry an umbrella. This strategy has always worked for me, so my wardrobe does not reflect improbable scenarios that go beyond my life experiences and action plans. If, one day, I were to embark on an extreme expedition, I would carefully strategize my textile choices, and the high-water resistance and breathable technology might become a priority.

1.3 Water-Repellency

Water-repellency in textiles relates to the surface tension of fibers, fabric structures and their relationship with that of contacting water molecules. This phenomenon can be easily observed in various objects such as water beading on lotus leaves or certain fabrics. The surface structure of, for example, lotus leaves creates a level of surface tension insufficiently strong to break the internal tension of water molecules, causing water droplets to maintain their shapes.

Cotton and Wool: Cotton fibers and fabrics have a high affinity for water due to the presence of the hydroxyl groups in cellulose. This causes water to penetrate and spread quickly through the materials. Wool, on the other hand, behaves differently because of its unique surface structure. While the protein in the wool also has a strong affinity for water, it is protected by the scales, as shown in *Fig.3*, on the fiber surface that provide a certain level of water repellency and resistance. However, once water penetrates through the gaps of the scales and wets the protein, it quickly spreads.

With their inherent affinity for water, fabrics made with cotton and wool may require chemical treatments to provide a desired level of water repellency, if used in textiles designed for direct contact with rain or snow.

Synthetic Materials: In contrast, tightly woven synthetic fabrics, even without chemical water-repellent treatments, can behave like lotus leaves to a certain degree, causing water droplets to bead up on the surface. This natural water repellency is closely related to the hydrophobic nature and smooth surface structure of most synthetic fibers.

Chemical Treatment: Per- and polyfluoroalkyl substances (PFAS) are applied to the weather-facing side of fabrics to create a high(er) degree of water repellency, preventing rain from wetting the surface. This treatment is commonly used on textiles made with natural fibers such as a woolen (brushed) winter coat, cotton jackets, etc. Another example is various outdoor jackets made with synthetic fibers. Although synthetic fibers naturally repel water as mentioned earlier, the chemical treatment boosts their performance to a level more suitable for avid outdoor activities.

The PFAS are organic compounds that come in different variations, such as C8, C6, and C4, with the numbers referring to the number of carbon atoms. Traditionally, the industry used the compounds with higher carbon numbers, like C8, also known as perfluorooctanoic acid (PFOA), because they offer superior water repellency and durability on textile materials compared to the compounds with fewer carbons.

Toxicity Recognition and Regulatory Works: In the late 1990s, several lawsuits were filed against DuPont by residents living near its Washington Works plant in Parkersburg, West Virginia, due to a PFOA contamination originated from the plant in the local drinking water. Subsequent research identified the PFOA as carcinogen, with a particular link to kidney and testicular cancers.

In 2006, the U.S. Environmental Protection Agency (EPA) and eight major companies that used these chemicals in various industrial applications agreed to the 2010/2015 PFOA Stewardship Program. The program's goal was to reduce PFOA emissions by 95% by 2010 and to work toward eliminating them by 2015. This initiative was successfully implemented, and by the mid-2010s, these companies had phased out the use of the PFOA in manufacturing consumer products and other intermediary materials such as fibers, yarns and fabrics.

In June 2020, the EPA issued the Significant New Use Rule (SNUR), requiring US manufacturers and importers to notify the agency before beginning or resuming the use of these chemicals for any significant new purposes.

Following the EPA's actions, the European Union also took major steps to regulate these chemicals. In 2019, many countries participating in the Stockholm Convention agreed to add the PFOA to the list of the Persistent Organic Pollutants (POPs).

Continued Use and Environmental Impacts: Despite these regulations, various PFAS chemicals are still used in many countries where environmental regulations are less strict. Additionally, significant loopholes exist in the regulatory actions of many countries. For instance, the EPA's action against the PFOA (C8) led to the use of other PFAS with shorter carbon chains, such as C6 (perfluorohexanoic acid, PFHxA) and C4 (perfluorobutane sulfonate, PFBS). However, concerns on the health impacts and the persistence of these chemicals in nature remain, as they belong to the same family of chemical groups.

Several studies have revealed the potential health impacts of the chemicals with fewer carbon atoms, such as thyroid hormone disruption, developmental toxicity, liver toxicity, and effects on the immune system at high doses. Even without these studies, scientific reasoning suggests that, while the impacts of C6 and C4 may be less severe than those of C8, they still exist and create proportionally harmful impacts.

These chemicals contaminate water either from manufacturing processes or laundering the textiles containing the chemicals, then spread globally through water streams. Moreover, the PFOAs applied to a large number of consumer and industrial products before the implementation of the environmental regulations still persist in the environment due to their extreme durability.

Combined with the ongoing use of these chemicals in many regions of the world, they will continue to impact the environment and human health for generations.

Consumer Awareness: As mentioned earlier regarding my approach to purchasing textile goods, the same principles apply when it comes to the water repellency feature.

I have often been caught in the rain before finding a shelter in my everyday clothing, and the feature could have provided some relief by keeping my body dry in such situations. With this, I recognize that the water repellency might be more useful for people like me in their city life than the water resistant and breathable technology. However, I still weigh my decisions carefully: These chemicals gradually wash out with each laundry cycle, eventually losing their effectiveness. The preference for C8 compounds over C6 and C4 largely stems from their superior durability on fibers and fabrics. Nevertheless, no matter how durable, this performance feature usually has a shorter lifespan than the textile item itself. In some cases, the repellency lasts only few washes, depending on wash conditions.

Justification: Given the limited duration of its effectiveness, as well as the associated costs and environmental impacts, it is hard for me to justify purchasing items employing this chemically treated feature. For these reasons, I actively seek out the products without any water-repellent treatment.

***

As the section title suggests, consumers are often "unfortunate" because they are frequently targeted by various stakeholders who promote a range of technical claims, making certain features seem relevant when, in reality, they may not be as beneficial as advertised. Additionally, consumers' choices are often limited by the decisions made by these stakeholders. For example, when shopping for a jacket, it can be difficult to find options that don't include unnecessary chemical treatments. Even if a shopper prefers to avoid these features, they may still end up purchasing a jacket with extensive chemical treatments, as these features are routinely applied to many textile products.

Unfortunately, consumers' limited understanding, coupled with a lack of interest in learning about the science behind these technologies, enables many organizations to prioritize performance claims over health and environmental concerns. This highlights the importance of consumer education. It is crucial for consumers to be informed about the practicality of certain performance features and to ask relevant questions about their environmental impacts. Given the importance of this issue, I explore it further in **Chapter V: Solutions**, particularly in the section titled **Education, Education, and Education**.

1.4 Artificially Conceptual Value (ACV)

An Artificially Conceptual Value (ACV) is a performance feature, that operates similarly to a placebo effect in medicine, but it can lead to unfavorable outcomes for the environmental sustainability.

Definition and Corporate Behaviors: To further define Artificially Conceptual Value (ACV), it refers to a technical feature that lacks scientific evidence to prove its effectiveness in real-life situations but makes people believe it works. Companies engaged in ACVs often introduce related scientific concepts to explain how their technologies theoretically function. Unfortunately, these explanations are frequently designed to quickly convince consumers who lack in-depth knowledge about the related science. This approach is highly effective because it seems logical and plausible on the surface.

Consequences: In the medical world, relying solely on placebo effect can worsen an illness over time, leading to serious consequences. For example, it is well known that Steve Jobs could have potentially saved his life by turning to modern medical treatments rather than relying on naturopathy for too long. The same logic applies to the environmental impacts of various ACVs. A scientifically accurate evaluation can help remove potentially harmful and unnecessary chemical treatments in the textile goods we consume for the good of environment.

***

Unfortunately, ACVs are widespread in the textile industry. I'm often astonished to see large, reputable, and respected companies engaging in such practices. In this subsection, we will review several examples of ACVs in textiles to help readers better understand this concept.

1.4.1 Heat Reflective Technology

A thin layer of a metallic silver pattern is applied to the inner lining fabric of winter jackets with the claim that it reflects the body heat back toward the wearer, conserving the heat and keeping him/her warmer than the person would be without this technology. This concept sounds logical and plausible to many, and as a result, the technology has become commercially successful in cold-climate regions where effective thermal technologies are essential for winter garments.

Heat Exchange: To understand the scientific foundation of this technology, it is essential to understand how the heat exchanges occur in the human body through five different mechanisms: conduction, convection, radiation, respiration, and evaporation, as illustrated in *Fig. 8*. These processes help prevent overheating by dissipating the heat generated from the food we consume, which raises body temperature, and helps maintain a constant temperature of approximately 36.5°C (97.7°F).

Fig.8 Heat Exchanges of Body

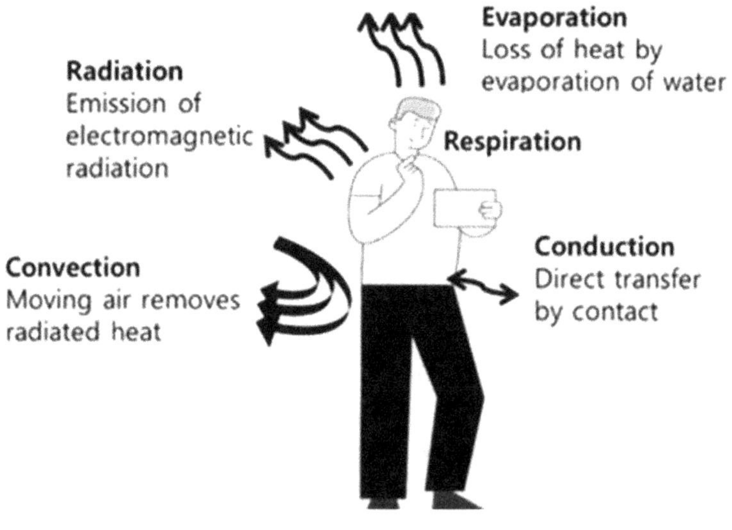

Evaporation
Loss of heat by evaporation of water

Radiation
Emission of electromagnetic radiation

Respiration

Convection
Moving air removes radiated heat

Conduction
Direct transfer by contact

The degree of the heat exchanges depends on several factors, including weather and individual body conditions, where some of these factors can be controlled, while others cannot. Without delving too deeply into the science of thermodynamics, simplified concepts are as follows:

Respiration, Convection, Conduction and Evaporation: Given a specific set of weather conditions and personal variables, our body heat escapes through respiration as we breathe. Wind removes the heat through convection, as though standing in front of a fan. A conductive heat loss happens when cold air molecules come into contact with the molecules warmed by our body heat, transferring the heat from the higher energy level of being warm to the lower energy level of the cold surroundings. An evaporative heat loss occurs through the sweat glands on the skin to release the heat through moisture, which evaporates and cools the body. Increased physical activity, such as running, generates more heat and moisture, leading to visible sweat and greater heat loss through this mechanism of evaporation.

Radiation: A radiative heat exchange, the focus of this heat-reflective technology, involves the transfer of the body heat through radiation. This type of heat exchange can be easily experienced by standing near a radiator in a room during winter; you can feel the warmth radiating upwards when the radiator is on. Similarly, someone engaged in intense physical activity, like vigorous running, can feel this radiative effect by placing her/his hand near the forehead. When the body needs to dissipate excess heat quickly, it radiates the heat through the skin, working in tandem with other mechanisms like heavy breathing and perspiration.

Mechanisms of Controllable Heat Exchanges: Of the five heat exchange mechanisms, three - radiation, respiration and evaporation - are mostly under the control of our body, while conduction and convection depend more heavily on weather conditions. These three "controllable" mechanisms are most active during a high physical activity in warm and humid conditions as the body needs to remove the heat more efficiently in those conditions.

Conversely, in a cold weather with a low activity level, the body attempts to conserve the heat by minimizing these three exchanges. In such cases, the respiration slows, the body's sweat pores shrink and close, and the skin works to prevent the heat from escaping, reducing the radiative heat loss to nearly negligible levels.

Hypothesis and Scientific Analysis: If the heat reflective technology were truly effective in real-life situations, one could wear nothing but a sheet of aluminum foil, seal it tightly to minimize convective heat loss from cold air and feel warmer than if they were nude. However, this setup would actually make the person feel colder due to another scientific principle: the aluminum has a much higher thermal conductivity than air, meaning it would transfer the cold energy more quickly. *Fig. 9* illustrates the thermal conductivity map of different materials.

Fig.9 Thermal Conductivity Chart

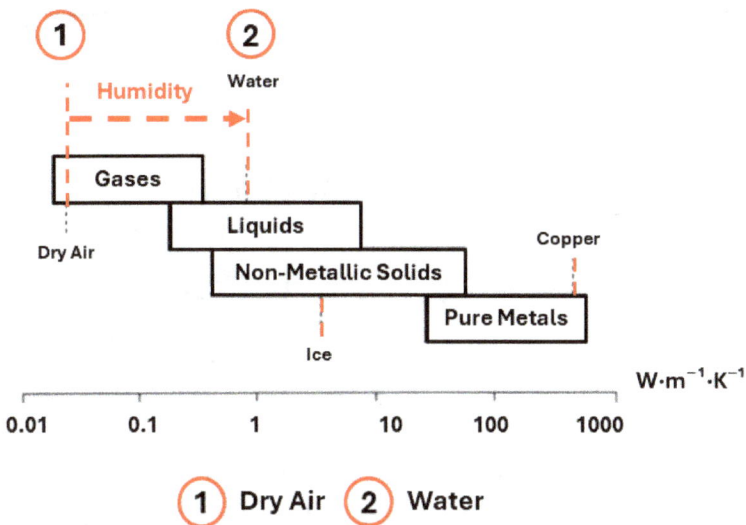

① Dry Air ② Water

Dry air is among the least conductive, while metals are among the most conductive.

Basics of Thermal Science: In the thermal science relating to textiles, a critical area of focus is the microclimate, the region between the body and the garment system, as shown in **Fig.10**. This microclimate traps warmer air generated by the body heat. The benefits of placing an aluminum sheet in the microclimate to prevent the radiative heat loss can be easily overshadowed by the rapid conductive heat loss caused by the metal's (or any other solid materials') high thermal conductivity.

Fig.10 Microclimate and Condensation

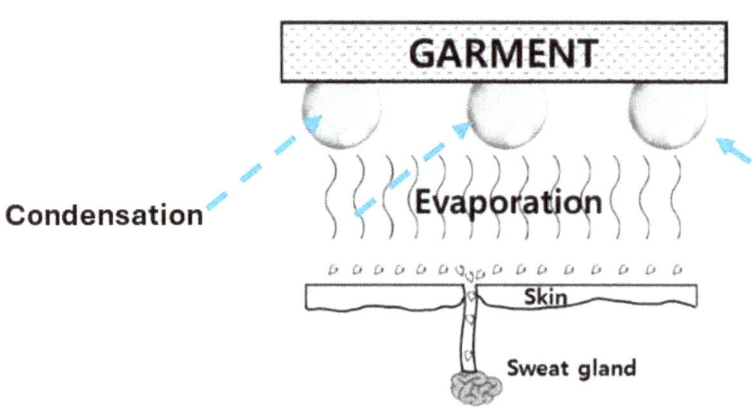

Test: A study was conducted to measure the effect of the heat reflective technology on the thermal efficiency of a textile structure typically used for winter garments. The study compared the CLO values of a conventional non-woven thermal insulation material with and without a lining fabric printed with the silver reflective dots, obtained from a specific brand's garments promoting this technology.

About CLO: The CLO is a metric used to measure the thermal efficiency of a material in the apparel industry, similar to the R-value used in the construction industry. A lower R-value indicates a less efficient thermal protection, leading to higher energy costs for heating and cooling of a house or building. Likewise, a higher CLO value in a textile material indicates a better thermal efficiency in conserving the body heat.

Test Environment: CLO tests are typically conducted in a controlled laboratory environment with a testing device, which includes a heat plate covered by the tested material, typically in a sheet form factor. The plate is heated to a certain temperature, for example, the body temperature of 36.5°C (97.7°F), and once the temperatures of the plate and the air above the material stabilize, the sensors located in the air above the material measure the heat energy trapped in the material by subtracting the temperature difference of the plate and the air, which is then expressed as a CLO value.

Results: The difference in the CLO values between the samples with and without the heat reflective liner fabric was less than 0.2 CLO, a range too small to be considered detectable by the general population. Given that the liner fabric hosting the reflective dots already has its own intrinsic CLO value, the impact of the reflective dots was negligible.

Real-World Scenarios: Under the premise of the heat-reflective technology, it might be more effective in cold weather when a person is engaged in high-intensity activities, causing the body to emit more radiative heat. However, in such scenarios, the body is already warm and doesn't need additional heat from the reflected energy. In fact, reflecting the heat back in these situations could do more harm than good by further raising the body temperature, increasing sweat levels in the microclimate, and potentially leading to hyperthermia, as heat exchanges occur more rapidly in moist air.

Importance of Thermal Technologies: Unlike highly specialized garments such as flame-resistant suits for firefighters or bulletproof vests for police officers, most everyday clothing is chosen for more mundane factors like appearance, materials, comfort, and price, rather than for its protective features. However, winter clothing is an exception, as it is essential for the health and safety of consumers in cold climates. While some may argue that a placebo effect in winter apparel is harmless if it provides a sense of warmth, in reality, relying on inadequate protection in cold weather can be dangerous and the potential harm should not be taken lightly.

Environmental Impacts: From an environmental and sustainability perspective, the printed patterns with different materials on the fabric surfaces introduce unnecessary impurities. While more details will be provided in **Chapter V: Solutions**, specifically in **Clean Recycling Initiative™** section, the success of recycling efforts, particularly with synthetics, relies heavily on the material purity and the foreign materials used in the heat reflective technology can hinder the efforts.

1.4.2 Phase-Change-Material (PCM) technology

Another performance feature whose value I find difficult to comprehend is the Phase Change Material (PCM) technology, commonly applied also in winter textiles. The PCM technology involves certain materials that change their phases from, for example, liquid to solid, and maintain a constant temperature during the phase change. For instance, when ice melts, it transitions from solid to liquid state, maintaining the constant temperature of 0°C (32°F) during the process. Similarly, water maintains the temperature of 100°C (212°F) during its phase change from liquid to vapor when boiling. A cooling gel used for stressed muscles maintains a consistent temperature while transitioning from semi-solid to liquid state on the skin.

Important Factors: The effectiveness of this PCM principle depends on several factors: the activation temperature, and the duration for which the material maintains a certain temperature. For example, to provide adequate protection in cold climates, the material must activate at a desired temperature and sustain it for a reasonable period. Therefore, evaluating the effectiveness of PCM in textiles requires an understanding of these factors. Unfortunately, many companies promoting PCM technologies often focus on marketing slogans like "*NASA used it for astronauts' suits on the moon*" rather than providing this crucial information.

Scientific Analysis: This misleads many people, who assume that, because the moon is extremely cold, this must be an effective technology. This is a typical example of how Artificially Conceptual Values (ACVs) can create misguided assumptions. However, for those with an analytical mindset, the issue is clear: the environmental conditions on the moon are vastly different from those on Earth, and the amount of PCM used in astronauts' suits is unknown. Our experience with ice gel packs for cooling sprained muscles provides insight into the importance of quantity. A gel pack of sufficient size and weight offers lasting relief, a concept which an astronaut's suit can easily accommodate in the effective use of PCM. However, if only a small amount were applied to a winter jacket as we cannot wear an attire as bulky and heavy as astronauts' suits, it would change its phase so quickly that the chemical application would offer virtually no benefit. Therefore, just because a technology works on the moon does not guarantee its effectiveness on Earth in real-life.

Environmental Impacts: Typically, a PCM applied on textiles is a chemical material derived from petroleum sources, with the paraffin wax being the most common type. The material consists of long-chain hydrocarbons with 20 to 40 carbon atoms. As discussed earlier, long-chain hydrocarbons are used in water repellent treatments for garments. Certain water repellent chemicals such as C8 are banned in many countries due to their carcinogenic properties and long persistence in nature. Although these PCMs are designed to adhere to textile materials for extended periods, they gradually wash off during laundering and regular wear, contaminating the environment.

When purchasing textile goods containing PCMs, it is essential to understand the specific technical details mentioned above. From a performance perspective, a purchase should only be justified if the performance metrics align with shopper's intended needs. In practical textile applications, I have yet to observe any benefit from PCM materials in real-world situations. In most cases, PCMs are combined with conventional thermal insulation materials because they are either insufficient on their own or not effective at all, which poses a question for its usefulness despite of the associated environmental harm.

1.4.3 Heat-Generating Technology

One prominent example of ACVs involves a heat generating technology, often marketed with various other performance benefits, including several health benefits such as blood circulation enhancement, negative ion emission for improved mood, etc.

Nature of ACVs and Analytical Skills of Consumers: After considering the earlier examples, readers might wonder why these apparent ACVs exist. In fact, there are many others that if I were to list and explain them all, this book would extend to additional hundreds of pages. Moreover, new ACVs will inevitably emerge as various organizations continue to find easy ways to appeal to consumers without making necessary efforts in truly effective technological advancements. I believe that the individuals and organizations making such claims often lack the depth of knowledge needed to fully understand the implications of their assertions. Unfortunately, this is a significant part of the dynamics of the consumer market today, once again highlighting the "unfortunate" nature of being a consumer.

Since it is impractical to cover all existing and potential future ACVs, the primary goal of this book is to help readers develop their analytical skills to better identify and recognize these misleading claims, an extendable skill for evaluating various environmental claims too.

Magical Technology?: Returning to the subject of the heat-generating technology, it can seem incredibly appealing, even almost magical, to believe that a piece of garment with the technology can offer such significant benefits, not only providing warmth but also improving blood circulation, mood enhancement, etc. Let us delve into the foundations of this technology.

Manufacturing Principles: Extremely small particles of various materials including ceramic, clay, carbon nanotube, etc., are added to and evenly distributed in a polymer tank where the polymer is in a molten state. Fibers are then extruded with these particles embedded in them. According to the companies which promote this technology, these particles alter the properties of the fibers with the proposed characteristics, creating the purported benefits. To substantiate these claims, they frequently use seemingly scientific data obtained from specifically designed tests such as the one explained below.

Test Methods and Results: A common testing method involves comparing two fabric samples: one containing the purportedly beneficial particles and another without. These fabrics are placed at the bottom of a test chamber with a high-wattage light source positioned directly above them. The specific test I observed utilized a 500-watt Kawasaki lamp placed approximately 20 ~ 30 centimeters above the fabric samples. Temperature probes are inserted beneath the tested fabrics to monitor the temperature changes.

With the test setup complete, the light source is activated, and the fabrics absorb the heat energy from the bulb. The fabric containing the particles is claimed to exhibit a faster rate of temperature increase compared to the control fabric without the particles. In the observed test, a temperature difference of 10°C (18°F) was achieved within 30 ~ 40 minutes after the initiation.

Bases of Claims: The observed temperature rise in the fabric containing the particles is interpreted as evidence of the increased energy absorption from the light source. The more energy the material absorbs, the more it emits in the desired forms of energy such as infrared-, negative ion, etc.

This premise suggests that the garments incorporating this technology would function similarly during real-life use, absorbing energy from sunlight or body heat and subsequently releasing it to generate the advertised benefits. Companies making such claims may present additional supporting information, drawing analogies to related natural phenomena that supposedly produce similar effects. Examples might include experiencing health benefits while sunbathing or walking in a forest, etc. However, the validity of these analogies and the scientific basis for the claimed functionalities require critical evaluation.

Interpretations and Truth

Several key issues undermine the credibility of such claims regarding the enhanced heat generation and other associated functionalities.

Energy Level: A high-wattage light bulb generates an extremely high surface temperature and emits a level of energy far exceeding what humans can tolerate. For example, a 500-watt Kawasaki lamp becomes extremely hot, with a surface temperature easily reaching 400 ~ 500°C (752 ~ 932°F). At this temperature, it is impossible to put skin within 20 ~ 30 cm (8 ~ 12 in) distance of the bulb without immediate burns. This illustrates that the test conditions used to support these claims are not realistic. Companies may argue that such tests are stress tests, intended to demonstrate the effect under extreme conditions and that the results are proportionally applicable to lower temperatures, such as those from the sun or human body heat. To test this hypothesis, I conducted an experiment.

Experiment: I took two pieces of fabric samples, one containing the particles and the other without, to the rooftop of a 10-story building on a hot sunny day, setting them up on the floor. The only principal difference from the laboratory test method was that the heat source was the sun instead of the 500-watt Kawasaki bulb. Although the conditions were not exactly same, my test was much closer to real-life scenarios. The rooftop was extremely hot due to the scorching sun, with an ambient temperature of 30°C (86°F) with no wind. After the setup, I monitored the temperature change on the two fabrics every 10 minutes. No difference was found between them for the duration of two hours, then I decided to stop.

Related Science: It is important to recognize that stress tests are typically performed to study a material's behavior under accelerated and extreme conditions. These tests are inherently limited when it comes to directly applying the results to real-world scenarios under normal conditions. Assuming that the outcomes observed under a 500-watt Kawasaki lamp will occur in real-life situations is overly simplistic and scientifically inaccurate. It is clear that even the test conditions I used are extreme for most people, such as being exposed for two hours under the scorching sun at an ambient temperature of 30°C (86°F). Moreover, the energy level from the sun in my test is unlikely to be encountered during winter, when this technology is more relevant, while the energy level our body generates is nowhere near that of the sunlight, let alone a 500-watt Kawasaki lamp.

Health Benefits?: Another perspective on this claim involves its purported health benefits. In some forms of oriental medicine, infrared lamps are used alongside acupuncture to enhance the blood circulation of a patient and promote healing; An infrared emitting lamp with a typical temperature range between 200 ~ 300°C (392 ~ 572°F) is directed at the treated area with a certain distance and limited duration of exposure, where needles are typically inserted into the skin, supposedly amplifying the therapeutic effects. However, this situation is analogous to the earlier example, where the heat energy from the infrared lamp significantly exceeds what people typically experience with sunlight and body heat.

Mechanics and Improbability: Examining the mechanics of this technology reveals further issues. The quantity of the particles used to create these fibers is limited because the molten polymer must be extruded through extremely small holes – more information on this extrusion process presented in **Chapter II: Synthesizing - Polymerization of Synthetic Fibers** section.

Adding the particles into a polymer melting tank is tricky because the small holes can be easily clogged up. Furthermore, the particles can settle to the bottom of the tank, which can create significant issues for continuous operations, such as fiber breakage and process failures. Typically, the particles do not exceed 10 ~ 15% of the weight of the molten polymer due to the concerns and process challenges.

Having had plenty of experiences in polymerization and extrusion processes in my career as textile engineer, I can attest that 10 ~ 15% is an extremely small quantity to create such magical benefits. Moreover, a large portion of the particles are encased within the fibers, surrounded by numerous polymer chains. For this technology to work, the sun's rays or the body's heat energy must penetrate the polymer structure, reach the particles, reflect enough energy, and then travel back through the structure to reach the skin. This scenario is, in my view, implausible even in science fiction.

***

If this technology truly had such remarkable benefits, one could simply bring a piece of rock into their living room and sit on it to gain the same effects, as the rocks naturally contain a much higher concentration of the substances claimed to provide these benefits. If it were that simple, this practice would already be widespread across the world.

Cautions and Environmental Impacts: Not to be persuaded by these types of ACVs, cautionary approaches are necessary. If a company presents test reports indicating medical effects, such as negative ion emission or far-infrared emission similar to the heat generation test results reviewed earlier, it is important to verify whether these tests were conducted under real-life conditions and whether the effects are proven to be truly beneficial for human health.

From an environmental perspective, the particles compromise the purity of the fibers and impede effective recycling, while the particles shed and contaminate, thereby negatively impacting the environment during manufacturing and consumption.

***

I've highlighted several examples of ACVs and performance features that, when fully considered, offer little or no practical benefit while potentially harming the environment. In fact, I regard some of these technologies and practices widely practiced in the industry as truly absurd.

Historically, profit-driven organizations have often prioritized profits over social responsibility and due diligence. Even in my own professional experiences, I have witnessed many instances of similarly natured corporate misconducts.

Consumers often place too much trust in organizations around us, assuming they have thoroughly done their homework and offer products grounded with valid sciences. Unfortunately, time and time again, it has been proven that this trust may be misplaced and exploited. This underscores the vital importance of consumer education, a topic I will explore further in **Chapter V: Solutions**, in the **Education, Education, and Education** section.

1.5 Other Important Textiles and Environmental Impacts

In this subsection, I will explore a couple of textile technologies that can either greatly benefit or harm consumers while also having significant environmental impacts.

1.5.1 Home Textile & Energy Savings

The modern world demands unprecedented amounts of energy, with technology advances such as artificial intelligence and electric vehicles further driving up the energy needs. Consequently, the importance of conserving energy has never been greater. With this pressing issue in mind, I introduce a simple, low-cost solution involving a textile application that can significantly reduce the worldwide energy consumption.

Construction and Energy Efficiency: Considerable efforts have been made to enhance the energy efficiency of our living environments, including homes, appliances, commercial buildings and maintenance practices, etc. Many governments offer incentive programs to encourage their citizens to adopt the materials and technologies that improve the energy efficiency of their living.

One such example is the effort to reduce the rate of heat exchange between the interior and exterior of establishments. As briefly mentioned along with the CLO, the thermal efficiency, expressed as R-value, is measured and optimized on construction materials. Furthermore, thermal imaging techniques are used to identify the areas of high heat exchange for necessary improvements.

Windows and Energy Loss: A defining trend in modern homes and buildings is the increased use of large windows for an aesthetic appeal and better views. Unfortunately, windows are inherently less efficient at preserving energy compared to walls, which are typically insulated with specialized materials. *Fig.11* shows an example of excessive heat loss through window areas.

Fig.11 Heat Loss through Windows

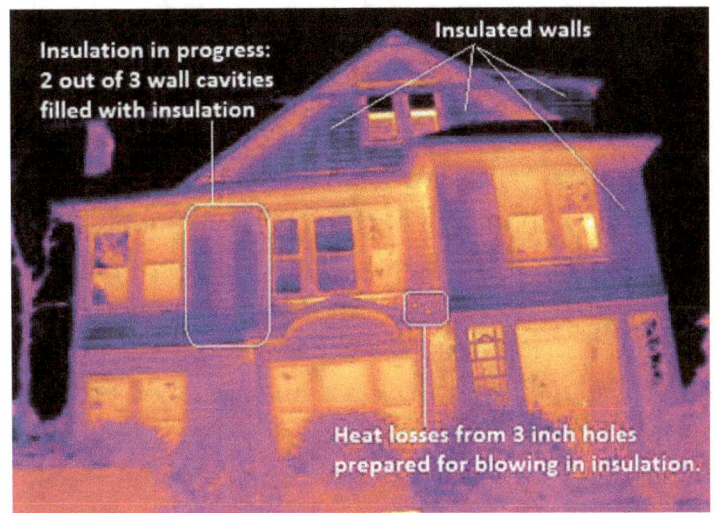

https://loe.org/shows/segmentprint.html?programID=12-P13-00046&segmentID=5

While many factors influence a building's energy efficiency, such as total window areas, efficiencies of air conditioning or heating systems, orientation of building and windows relative to the sun, integrity of window frames and seals, etc., some studies have shown that a significant portion of energy, up to 70 ~ 90%, can be lost through the windows in thermally inefficient buildings.

Window Tint: Tinted glass windows can reduce heat gain during summer by blocking parts of sunlight, leading to lower cooling costs. However, this approach has its limitations. Many people value window transparency and are hesitant to darken their windows too much, which reduces potential energy savings. Additionally, tinted windows can block beneficial heat gain in winter, increasing heating demand.

Thermal Break: Some newer window frames are designed with two or more layers of glass, separated by gaps that act as a thermal break, helping reduce energy loss in winter and heat gain in summer. While this solution can improve energy efficiency, it is often costly, potentially requiring thousands of dollars to replace existing window frames, even for small homes. For large commercial buildings with older window frames, retrofitting can be nearly impossible. Although this technology can reduce energy bills in both summer and winter, its effectiveness is somewhat limited by the laws of thermodynamics: larger air pockets between the glass panes transfer energy more efficiently than smaller, compartmentalized pockets, making it less thermally efficient overall despite the high costs.

Other Means: Some people turn to solutions like window films or bubble wrap to save energy, particularly during winter in colder climates. However, these options require considerable effort for seasonal installation and removal with annually recurring costs while incurring environmental burden from the used and disposed materials.

***

Although the measures above can help conserve energy, combining them with, or even relying solely on, "thermally insulated curtains" can greatly enhance the thermal efficiency of homes and commercial buildings.

Thermally Insulated Curtains

The thermally insulated curtains are a cost-effective and simple solution suitable for any homes or commercial buildings, regardless of the age of construction or window size and location. These curtains are made with specifically designed synthetic non-woven fabrics, which are engineered to provide optimal thermal efficiencies while using a minimal quantity of textile materials.

During the manufacturing process of such non-woven fabrics, numerous tiny air pockets are created, forming an effective thermal barrier year-round. In summer, these curtains minimize heat gain from sunlight, creating a cooling effect, while in winter, they minimize heat loss from heating systems. Although heavy blackout curtains can provide similar benefits, optimally engineered thermally insulated curtains offer superior thermal efficiency, leading to a better energy conservation while using less textile materials.

These thermally efficient curtains can effectively address all forms of the heat exchanges in a building, namely conduction, convection and radiation. In general, the effects they create are far superior to any other methods mentioned earlier.

Commercial Availability: Despite their significant benefits, affordability, and flexibility, these thermally insulated curtains are difficult to find in the market. A USA retailer briefly offered them many years ago, but the product line was discontinued due to a low popularity and several logistical challenges associated with the product for the retailer. For example, these curtains took up considerably more spaces in their retail stores and warehouses compared to their non-insulated counterparts. Although this difference is negligible once installed in homes and buildings, the increased retail space requirement and associated challenges combined with a low popularity among consumers likely influenced the retailer's decision to discontinue the product.

Consumer Awareness: In my view, two main issues contributed to the unsuccessful market deployment of these curtains at the time.

Firstly, had the consumers been made aware of the potential savings with hundreds or even thousands of dollars in electricity bills and significant environmental benefits, many more would have purchased the product. A comparable example is the widespread adoption of solar panels on homes and buildings, where homeowners are willing to invest thousands or tens of thousands of dollars while enduring significant renovation work during the construction of their living quarters.

Secondly, the increased storage spaces and handling costs for the retailer could have been easily offset by higher price tags. If consumers had understood the clear benefits, many would have been willing to pay a premium; Based on my industry experience and knowledge of textile goods' cost structures, the additional cost to consumers for these products would likely be only a few dollars. Furthermore, the logistical challenges could have been addressed with a vacuum packaging, reducing the product's volume and making it comparable to, or even smaller than, other heavy blackout curtains. While vacuum packaging might slightly reduce the thermal efficiency of the product, a careful engineering can minimize this impact.

Based on this analysis, I believe the failure of the retailer's thermally insulated curtain program was a result of poor execution in promoting the clear benefits both economically and environmentally. If any retailers, curtain brands, or related businesses are reading this and interested in exploring this concept, I would be more than happy to assist with product engineering, logistics, and even some marketing ideas.

1.5.2 Harmful Face Masks for Health and Environment

Even well before the COVID-19 pandemic, many people in Asia had recognized the importance of wearing face masks when facing respiratory viral infections due to several previous health crises like the avian flu, the H1N1 influenza, the MERS, etc. Additionally, masks were commonly worn as a protection against severe air pollutions in heavily industrialized regions. So, when the COVID-19 emerged, people instinctively reached for masks in their closets or from local stores.

For many, it was surprising when Dr. Anthony Fauci, the head of the National Institute of Allergy and Infectious Diseases (NIAID) in US, initially stated that masks were unnecessary for the general public. He emphasized the importance of reserving masks for healthcare workers and suggested that while masks might offer some protection, they were not essential for those who were not infected or caring for the infected.

Then, several weeks later, the guidance was reversed, and the public was urged to wear masks during all social interactions. While there is an ongoing debate, albeit by some minority opinions, about the effectiveness of face masks in preventing the spread of respiratory viral infections and, also acknowledging my lack of expertise in the medical science, I refrain from drawing any conclusions about whether wearing face masks is right or wrong in this discussion. However, based on the conventional understanding promoted by the authorities around the world after a short period following the breakout and other similar viral outbreaks of the past, the initial response from Dr. Fauci appears to have been mistaken.

Importance of Developing Analytical Skills: In my view, prioritizing healthcare workers' protection was something most of our society would have readily supported. However, providing accurate information to the public is equally critical in the efforts to mitigate the crises of this nature. The initial lack of understanding in some countries about the asymptomatic transmission of the COVID-19 stood in a stark contrast to the immediate mask-wearing response in many Asian countries.

This incident highlights that even top authorities can make mistakes. In fact, there have been many other examples of such occurrences, some of which will be discussed later in this chapter and in **Chapter V: Solutions**, specifically in the **United Efforts** section. These examples remind us that blindly following authorities may not always lead to the best outcomes, especially when it comes to matters as crucial as our health and safety.

While I do not aim to promote conspiracy theories or encourage those who spread misinformation and provoke unwarranted fear, we must develop analytical skills needed to recognize flawed guidance. In such situations, it is crucial to approach the issues with a scientific and analytical mindset and, at times, rely on common sense. While expert advice is invaluable, applying critical thinking and questioning authority on questionable guidance is key to managing our lives more effectively.

Copper Particle Masks

The lengthy prelude to this subsection serves to alert people about a health product that poses significant risks.

Introduction of Dangerous Mask: Recently, I was in an online meeting with an Asian manufacturer regarding a specific technology I was seeking. During the call, I learned that the company was involved in various other businesses, including face masks. After completing the discussions on the planed subject, the manufacturer began promoting a specific mask product, which featured copper nanoparticles. As they showcased a sample on the camera, I noticed a diluted yellowish tint evenly distributed across the mask, indicating the presence of copper in small particle forms.

From a particular research project during my tenure as research and development engineer in South Korea, I gained considerable insights about the concept behind this product: the copper particles are intended to kill microorganisms by releasing electrically charged ions that destroy viruses and bacterial cells in contact. Silver has similar properties, but it is much more expensive. The manufacturer claimed that these masks were quite popular with customers in many countries including USA, Japan and many countries in Europe.

Significant Concerns and Flag: However, my experience in the project in South Korea immediately raised serious concerns on this mask product. Metal particles, such as copper and silver, cannot be permanently fixed to fibers and fabric structures. In fact, a significant quantity of these particles is released into the air during use, and the release quantity depends on how they were impregnated into the material structures during manufacturing.

Research Project: In the research project, I was tasked with developing acrylic fibers impregnated with copper particles to reduce static electricity that people experience when putting on or removing acrylic sweaters. The idea was that the metallic ions in copper would help dissipate the stagnant electricity. From that experience, I learned that, despite my best efforts to keep the process clean, copper particles contaminated the entire laboratory surfaces. When I swiped my fingers across any surface, I observed a greenish-yellowish tint, typical of the copper particles in high concentration, on my skin.

Danger and Potential Health Impacts: Placing such particles in the path of our breath is dangerous. It is not difficult to imagine that these particles can be easily inhaled through the mouth and nose while breathing. Once inhaled, they can reach organs and bloodstreams in the body. For instance, asbestos, now banned in many countries after extensive research, causes lung cancer because its fibrous particles scratch and damage lung tissues. Metals like copper, being potentially harder than the asbestos, could inflict even more damage inside the body. Additionally, metals can trigger other unknown reactions due to their heightened chemical reactivities.

After my meeting with the manufacturer, I searched the ChatGPT, which resulted in the following description: *"During the COVID-19 pandemic, the FDA issued Emergency Use Authorizations (EUAs) to facilitate the availability of various types of face masks, including those with antimicrobial materials like copper and silver, to help prevent the spread of the virus."*

Although I immediately advised the manufacturer against selling the mask product during the meeting, they likely believed that the FDA's approval carried higher authority and validity than my anecdote and assessment for the health risks.

Variables and Environmental Impacts: If a mask is made with copper or silver in a strong filament yarn form factor, it could serve its intended purpose as antimicrobial with a lower health risk, as the continuous filaments are much less likely to shed particles during use. However, the masks made with copper or silver in such form factors as particles or fibers significantly increase the risk of the particle inhalation. In my research project, I embedded fine copper particles into molten acrylic polymer in polymerization process and then extruded it into fibers. This process is one of the most secure methods of embedding fine particles into various forms of polymer structures. However, even with this secure method, the resulting fibers still contained considerable quantities of copper particles on their surfaces, which easily transferred onto my skin when touched.

Readers, please do not wear the face masks made with any particles regardless of its kind. Inhaling the particles can pose more harm than any potential benefit.

Environmentally, small particles of metals with their inherently heightened chemical reactivity will persist in the environment and create significantly harmful effects on all living organisms.

Section 2 "Fast Fashion"

Although this topic aligns with the theme of the previous section, **Unfortunate Consumers**, I felt it necessary to create a dedicated section due to the widespread misconceptions surrounding it and its importance in key environmental initiatives.

"The term, "fast fashion" was first popularized in a New York Times article in 1989 to describe retail store Zara's first opening in the United States" – CNN

https://www.cnn.com/style/what-is-fast fashion-sustainable-fashion/index.html

Since then, the concept has rapidly gained global attention and is now frequently used as a symbol of various environmental issues associated with textile goods and organizations, often based on unscientific and vague notions.

Disagreement: The term "fast fashion" is encapsulated in the quotation marks throughout this book, reflecting my disagreement regarding its conventional usage and definition. This reservation is rooted in my background in textile engineering and decades of industry experience, which provide me with a deeper understanding of the complexities of textile production, quality, and market dynamics than most in the public, media, environmental advocacy groups and activists, including the New York Times reporter who created it.

Varying Perceptions: The ambiguity surrounding the concept of "fast fashion" is evident in the varied public perceptions of different apparel brands in the market. Some consumers may categorize a particular brand as "fast fashion," while others may not, highlighting the subjective nature of this classification. These divergent opinions often arise from the differing interpretations of various aspects of business operations, which may not necessarily relate to the quality or durability of the goods categorized as "fast fashion".

As a habit of a person, deeply involved in the textile industry, I often conduct informal surveys, asking people around me on various subjects related to textiles. One such example is to ask them to classify well-known brands such as Lululemon, Adidas, Nike, etc., as either "fast fashion" or not. The responses I receive are notably diverse, with no clear consensus emerging for any particular brand. This variability in people's perception extends even to some brands, which are considered as "luxury" brands by a large portion of the public, primarily due to the high retail values of their merchandises, surprisingly classified by some as "fast fashion".

Uneducated Notions: Given the vagueness, the public might recognize that the original coining of the term "fast fashion" in the 1989 New York Times article may have been based on uninformed notions by the author, rather than a clear and scientific understanding of the complexities of textile quality.

In light of this, I present several points for consideration concerning this situation in the following discussions. These deeper examinations will reflect my view of the term as more of a social phenomenon than the accurate measure of textile quality and durability, underscoring the importance of a more nuanced understanding of textile materials and their environmental impacts.

2.1 Ambiguity in Definition

From the perspective of the New York Times reporter, all garments in Zara's stores would be categorized as "fast fashion," regardless of material choices, durability, or other factors related to the quality. This label comes with several associated perceptions and environmental concerns including:

Perceptions

- *"Fast fashion" goods are produced quickly and have shorter lifecycles due to poor quality.*

- *"Fast fashion" goods are made with cheap materials, a notion frequently associated with synthetic fibers*

- *"Fast fashion" goods persist in nature for too long, a notion once again associated with synthetic fibers*

- *"Fast fashion" goods do not last long in consumption*

- *"Fast fashion" goods create microplastic pollution from synthetic fibres*

- *"Fast fashion" goods lead to overconsumption*

- *"Fast fashion" increases carbon footprint due to reliance on air shipments rather than ocean transport*

Furthermore, the opposite concept is often described as "slow fashion" and represent more favourable images consumers form towards certain materials, merchandises, brands, forms of operations, etc.

Expanded Interpretations: In addition, the term "fast fashion" has acquired connotations beyond its original meaning, often encompassing broader social aspects of a company's operations. For instance, some individuals automatically associate overseas production utilizing lower labor costs with "fast fashion", regardless of the actual quality of the goods produced.

Conversely, the varying interpretation of "fast fashion" appears to blur the boundaries between the product-specific characteristics in quality measurements and the broader concept of ESG (Environmental, Social, and Governance) practices within a company's overall operations. While some may argue that this broader definition provides a more comprehensive representation of a company or brand, there are drawbacks to such an expansive interpretation, particularly, in the perspectives of making more effective environmental efforts.

Mixed Values and Discrepancies: For instance, the quality, durability, and speed of factory-to-market for a particular product can be managed through specific operational strategies, which may be distinct from an organization's commitment to social and environmental responsibilities. If a company aims to align with consumers' notion that synthetic fiber products are inferior in quality and longevity, they might simply substitute natural fibers for synthetic ones. The perceived improvement in the longevity from the switch would lead equally to the perceived improvement in sustainability and this action can be executed independently of the company's efforts to fulfill its other responsibilities.

Combining these diverse concepts under a single term, therefore, may impede the public's ability to accurately evaluate matters related to the "fast fashion" and its overall environmental impacts. The following subsections will explore how ambiguous notions of the "fast fashion" can potentially lead to unintended consequences for environmental and sustainability efforts.

2.2 Manufacturing Factors – Material, Quality and Cost

This subsection begins by exploring the concept of "Quality", a term often misunderstood in the context of "fast fashion."

Definition of Quality – Oxford Languages

- *the standard of something as measured against other things of a similar kind; the degree of excellence of something.*

- *a distinctive attribute or characteristic possessed by someone or something.*

The definition emphasizes the "measurable" attributes or characteristics, which contrasts with ambiguity often associated with the products or brands labeled as "fast fashion". Therefore, using specific metrics commonly applied in evaluating textile quality provides a more scientific basis for assessing the true quality of the materials and goods associated with "fast fashion" and "slow fashion".

While "quality" can refer to many different factors, the following analyses focus on the material aspects of natural and synthetic fibers, as well as their derivatives like yarns, fabrics, and finished products. For instance, if a seamline breaks, it is more likely due to poor workmanship or the weakness of the sewing thread, rather than the fabric's strength. Although a consumer may interpret this as a sign of "poor" quality, this discussion distinguishes the scientifically measurable durability of the fabric from other potential causes of failure. Here, I present these quality factors with a focus on the material considerations.

Strength: Synthetic fibers are generally five to ten times stronger than natural fibers - A detailed analysis of this topic will be presented in the section, ***3.2.1 Strength Factor and Dust Quantity***, with supporting scientific data. This strength advantage means that products made from synthetic fibers are better able to withstand forces such as abrasion, tensile stress, tear strength, stretching, etc., key factors that contribute to the failure of textile goods. The higher strength of synthetic fibers directly enhances the durability of the resulting materials, including yarns, fabrics, and finished products, compared to those made from natural fibers.

Durability Factors: While this is a straightforward scientific fact, it can become more complex when different care methods are applied to different items, as previously reviewed in this chapter. In a brief recap, home laundry involves water usage and significant physical forces from the washing mechanisms, leading to accelerated material fatigue and degrade whereas drycleaning does not involve such forces. Additionally, interactions between the chemicals contained in textiles, water and detergents can increase the risk of color fading. Although synthetic materials generally withstand these physical and chemical forces exceptionally well, comparing home-washed items with dry-cleaned ones is not an accurate measure of their overall quality and durability.

Contradictory Perceptions: Furthermore, contradictory perceptions exist regarding "fast fashion" goods: *"They do not last long in consumption, but last too long in nature."*

Scientifically, if a material lasts long in nature, it typically reflects a quality that also offers longevity in consumption. Therefore, an analytical mind can draw a conclusion that the perception, contradictory to the underlying science, may be created by different metrics such as varying measuring sticks with subjective quality references. Conversely, there is a significant gap in the public perception of quality and designations of certain brands as "fast fashion" or "slow fashion."

***

I recognize that many readers may not yet grasp the full extent of the scope presented so far in this discussion. It is natural as textile quality and manufacturing are more complex than most people think. With the tangible examples in the following discussions, the scope will become clearer.

Complex Manufacturing and Market Structures

Throughout my career as textile engineer, I have visited countless factories involved in every stage of textile manufacturing, from raw materials to finished goods. These experiences have shown me that it is not uncommon to find a single factory producing goods for both "slow fashion" and "fast fashion" brands simultaneously, using identical raw materials, craftsmanship, and processes.

Similar Qualities with Different Prices: These goods are essentially of similar quality, if not identical, but their price tags can differ considerably based on the brands in question and their retail strategies. For instance, one brand may sell its products in an upscale storefront, while another sells online with large differences in the prices between the two. Without knowing the origins of these goods, many consumers may distinguish them as either "fast fashion" or "slow fashion", "good quality" or "poor quality". This example illustrates the complexities of manufacturing and market structures which most consumers do not fully understand, yet it can lead to improper judgements on the quality of goods, leading to actions which impact environmental sustainability.

Different Qualities with Different Prices: Different scenarios do exist, which correspond more to consumers' general expectations for a higher quality in relation with a higher price in retail. For instance, a brand implements rigorous quality control programs to ensure that their products respect safety and quality standards for its workers and consumers. This may include avoiding harmful chemicals, which can be otherwise used to enhance visual appeal or reduce production costs, among many other operational measures taken by brands and manufacturers. Such approach requires a system with necessary resources, such as experienced personnel and effective operational structures. This responsible approach adds to the cost structure of the organization, which is naturally reflected in the pricing of their products.

However, it is essential to recognize that certain operational systems and cost structures, while critical, should not be confused with the material quality of the products themselves in the context of "fast fashion" vs. "slow fashion" and its associated misperceptions described earlier. This distinction leads to a broader discussion on the nature of material quality and cost as follow.

Materials Matter: To illustrate the perspective of material choices, associated with their impacts on quality and costs, I use a scenario involving a brand using silk material for producing a dress instead of polyester, leading to a significant cost increase. Many consumers would prefer an expensive silk dress if they could afford it, not considering it as a "fast fashion" item. However, in terms of the "measurable" strength and durability, an equivalent dress made with polyester offers much higher strength, thus would last longer under identical care conditions. The silk dress could be ruined in a single heavy home laundry cycle, whereas the polyester dress would endure repeated washes.

2.3 Retail Price Factors

Among consumers, there is a common belief expressed by phrases like "*you get what you pay for*", suggesting that higher prices equate better quality. This subsection explores the disconnect between this perception and the actual impact of material costs on the retail prices of various textile goods.

Cost Bases: Revisiting the example of silk and polyester dresses, it is evident that a portion of the cost is tied to materials, such as the silk yarn for the silk dress and the polyester yarn for the polyester counterpart. However, numerous other factors contribute also to the final retail price. In fact, several other key factors exert a stronger influence on the cost structures, which in turn affect retail prices. As mentioned earlier in the discussion on **Different Qualities with Different Prices**, some costs are related to quality assurance. Of all associated factors in manufacturing, however, labor costs typically represent the largest expense, often far exceeding the cost of materials.

Other significant factors are non-manufacturing related, such as the point of sale (POS) structures and administrative functions. It is widely understood within the industry that a substantial portion of the costs associated with retail products comes from administrative operations, often far exceeding manufacturing costs, including materials. These administrative costs include employee and executive salaries, benefits, office space, storefronts, inventory storage in major cities, and marketing expenses for advertising campaigns.

While retail prices are shaped by complex operational structures that cannot be fully understood without examining a company's cost base, certain patterns are well known in the industry. For instance, special sales events like Boxing Day and Black Friday, offering, for example, 50% discounts, often still yield profits for the sellers. End-of-season sales with discounts like 70% may bring companies to a break-even point or close to it, covering both manufacturing and administrative costs.

By applying this logic (material cost < labor cost < administrative cost + 50% profit, for example), it becomes clear that material costs represent only a small fraction of the retail prices of the most textile goods we purchase.

Pricing Strategies: Additionally, pricing strategies vary across companies. Some brands may opt for lower sales volume with higher price points to maintain a desired brand value. Conversely, other companies may prioritize maximizing profits by selling larger quantities at lower margins. These strategies are often influenced by factors such as existing brand values, target markets, demographics, among many others.

Disparities – Quality and Price: Regardless of the rationale behind such price variations, a significant gap between raw material costs and retail prices often exists across both expensive and affordable textile goods. This disparity can result in consumers paying a premium for perceived brand value or administrative overhead rather than the inherent quality of the products.

Personally, I prioritize the value proposition of the goods I purchase rather than assuming a direct correlation between brand recognition and product quality. This approach involves a comprehensive evaluation that considers price, quality, durability, maintenance costs, and overall performance.

For instance, a suit I purchased from Zara, the brand initially labeled as "fast fashion" in the New York Times article, has remained in excellent condition for over 15 years. Contrary to the reporter's perception, I see no reason to associate it with the negative connotations of "fast fashion" in terms of its quality. Despite avoiding highly recognized brands and expensive prices, I rarely fail in my purchasing decisions, even considering the environmental sustainability aspects of the items I buy.

This book aims to provide readers with the knowledge they need to make informed decisions in their textile purchases, including understanding how their choices impact the environment.

2.4 Misperceptions and Overconsumptions

Apart from the quality and longevity perspectives presented earlier, labeling certain brands as "fast fashion" can inadvertently create misperceptions, leading to undesired environmental consequences by influencing consumers' overall consumption habits of textile goods.

Emotional Values: Most people, including myself, have certain belongings that they feel more attached to than others. They can be expensive items like houses and vehicles, or special gifts attached with precious memories. Conversely, special attractions and perceptual values can certainly exist in textile goods. Personally, I value clothing that is comfortable to wear and easy to maintain, which stems from my background as textile engineer. This perspective may differ from that of the general public, who might not assess textile goods through the same level of technical lens as me. However, like others, I also attach emotional value to certain items, especially those associated with meaningful moments or gifted by special people. These items are often more cherished and cared for due to the emotional connections they hold.

Lower Values: On the other hand, some items may be regarded as having lower value by many. For example, an inexpensive synthetic shirt displayed in a store associated with "fast fashion" may lead a shopper to make an impulsive purchase. In this situation, the shopper may develop preconceived notions about the item's quality and durability, often aligning with the common perception that *"fast fashion" products have shorter lifespan"*.

Consumption Behaviors: Relating to emotional values and purchasing decisions, it is natural for most consumers to think more carefully when buying higher-priced items. The more expensive the product, the more planning and consideration go into the decision, which in turn creates an emotional value in itself. For this reason, people often vividly remember where and when they bought an expensive silk dress, while they may not have such clear memories of purchasing an inexpensive shirt on impulse.

Differences of Fates: Perceptions also influence how people treat items that show signs of wear. If someone believes an item isn't meant to last, even a minor flaw, like a loose button or a broken seam, can prompt them to dispose of or donate the item rather than repair it. In times when materials were scarce, such issues would have been resolved with simple repairs, no question asked. Today's picture in wealthy countries, however, presents a stark contrast. Large quantities of textile items placed in community recycling bins or donated to charity groups often require only minor repairs. It is evident that the individuals who disposed of these items had little or no emotional attachment and valued them too little to retain them with small fixes.

***

In summary, labeling textiles as "fast fashion" can create a perception of lower value in consumers' minds, which may not accurately reflect the real quality or the longevity of the goods in question. This perception shapes the fate of these items, often leading to premature disposal in landfills or incinerators and subsequently driving the consumption of other textile goods as replacements.

Section 3   Microplastic vs. Natural Fiber Dust

Dust, by definition, consists of small particles detached from their original body mass, either from natural or synthetic materials, due to various physical or chemical impacts, then spread primarily through air. Many individuals suffer from allergies caused by pollens during flowering seasons. Pollens are an example of dust originating from nature, illustrating that not all natural sources are harmless. The impacts of dust particles on health and environment can vary widely depending on two main factors: chemical compositions and quantity.

Since the invention of the plastics over a century ago, these materials have become deeply embedded in our daily lives. Due to their thin and long form factors, synthetic fibers have been a major source of microplastics for more than 100 years, shedding dust particles during manufacturing, normal wear and washing.

This contamination has led to growing concerns about microplastics in virtually all our food sources, and rightfully so. It is well known that microplastics are present in almost everything we eat and drink, even in unexpected places like inside meat, fish, and vegetables. These tiny particles are so small that they even float in the air we breathe.

In this discussion, I delve into the nature of the dust particles from textiles and their environmental and health impacts.

Natural Fibre Dust and...: We would be living in a rather strange, and almost magical, world if only synthetic fibers shed dust (microplastics), especially considering that natural fibers are generally much weaker than synthetic ones - a point that will be explored in more depth in subsequent discussions later in this section. Yet, the dust from the natural fibers is largely overlooked. As a result, many readers may be unaware of its nature and impacts.

***

To address this topic on a scientifically fair basis, I will examine the sources and impacts of the dust from both synthetic and natural fibers, beginning with the scientific understanding on the microplastic.

3.1 Microplastics and Synthetic Fibers

Microplastics are generated from plastic sources through physical impacts during manufacturing and consumption. A common misconception is that microplastics mainly originate from the obvious plastic materials we encounter routinely in our daily lives like bottles and bags. While these sources do contribute, one of the most significant sources of the microplastics is the synthetic fibers in textile goods. This is due to their extreme length directionality and inherent weakness along the length side.

Forces for Microplastic Creations against Polymer Formation Structures: Scientifically, there are 13 different physical forces that can cause materials to break, and in the case of plastics, this breakage often results in the creation of what is commonly referred to as "microplastic".

These forces include tensile, compressive, shear, bending, abrasive, and frictional forces, among others.

When it comes to the structural integrity, which plays a crucial role in determining the amount of dust produced, plastic bottles and bags offer much greater resistance to physical impacts compared to synthetic fibers. This is due to their more favorable formation structures and form factors, which become apparent when examining their surface structures. Under extreme magnification, modern microscopes reveal that the polymer chains in plastic bottles or bags are tightly intertwined, providing strong structural integrity, as shown in *Fig.12*.

Fig.12 Plastic Formation in Surace Structure

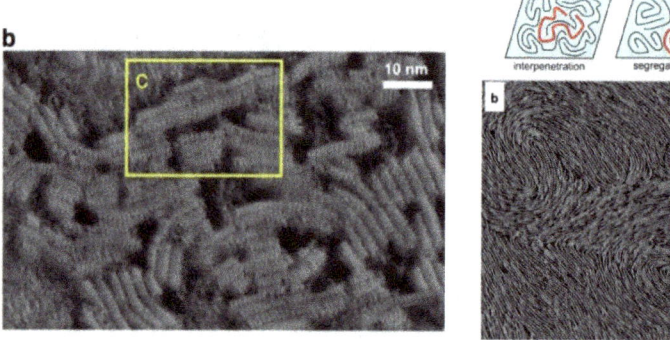

https://www.nature.com/articles/pj201567

In this robust structure, the polymer chains are arranged in random directions, making the material inherently resistant from all directions of the physical forces mentioned earlier. Weaknesses arises, however, when plastics are formed into thin sheets, like grocery bags, which are more susceptible to stretching, tearing and puncturing compared to thicker plastics, such as food containers.

Despite this, thin plastic bags still demonstrate an impressive strength-to-weight ratio, often capable of holding thousands of times their own weight (typically a 1:1,000 to 1:2,000 weight ratio). This combination of strength and lightness has contributed to the widespread popularity of plastic bags and bottles, while significant environmental concerns surrounding them have grown.

Microplastics from Synthetic Fibers: In contrast, synthetic fibers are far more susceptible to breakage due to the extreme directionality created by the long and thin form factors. The difference in strength becomes evident when comparing the two: while a thin plastic bag weighing 5 grams can hold 5 ~ 10 kg (11 ~ 22 lbs) of groceries, synthetic fibers can break under just a few grams of weight, which explains why a large number of them are bundled together and twisted into a yarn in part to enhance the strength – more on this in **Chapter II: Yarn-Making** discussion.

While the resulting yarn becomes considerably stronger, the comprising fibers remain weak. As a result, mundane everyday activities such as putting on or taking off garments, friction with other objects, or laundering can easily generate enough force to produce microplastics.

A study by a team at Plymouth University highlights the significant amount of microplastics released during laundering. They reported that an average washing load of 6 kg of laundry could release an estimated 137,951 fibers from polyester-cotton blend fabric, 496,030 fibers from 100% polyester, and 728,789 fibers from 100% acrylic.

Cautionary Analysis: While the data clearly indicates the generation of a large quantity of microplastics from a single laundry load, I approach these findings with caution. Three key considerations warrant this careful analysis: 1) the methods used to capture microplastics, as even the most advanced filtering systems cannot capture all microplastics due to their extremely small and lightweight nature, 2) the difficulty in distinguishing between the dust from natural and synthetic materials, especially in the blended fabric of the polyester-cotton mixes; and 3) the lack of information on the quantity of the dust from the cotton in the blended fabric.

3.2 Dust of Natural Fibers

Similar to synthetic fibers, natural fibers exhibit extreme directionality with the thin and long form factors. The same physical forces that generate microplastics from the synthetic fibers, such as tensile, shear, and frictional forces, also apply to the natural fibers. In fact, these forces can have much greater impacts on the natural fibers, leading to the production of significantly larger quantities of dust compared to their synthetic equivalents under the same amount of physical force.

Contrary to the popular belief, the dust from the natural fibers can have more harmful environmental impacts due to the larger amounts of residual chemicals and their higher toxicity levels. Toxicity aspect related to different textile materials, a central theme of this book, requires detailed analysis and a comprehensive understanding, as it plays a crucial role in our environmental management. As such, this topic will remain a focal point and will be discussed in depth throughout the relevant sections of the book.

3.2.1 Strength Factor and Dust Quantity

To comprehend the quantity of dust generated from textile materials in both natural and synthetic compositions, it is essential to understand the strength of each type of fiber and how it is measured. In this discussion, we employ the breaking energy analysis method.

Industry Specifics (1) – Start

Test Method: Each fiber strand is affixed to two gripping mechanisms, one at the top and one at the bottom, within the test equipment, illustrated in *Fig.13*. The top gripping head then pulls the fiber strand in the length direction until it breaks. During this process, the sensor attached to the moving head measures the breaking force and the distance the fiber travels until breakage. This method, commonly referred to as "tensile test", is also applicable to various other intermediary textile materials such as yarns and fabrics. Beyond the textile applications, it is often used in such materials as films, steel, etc., where tensile strength is a critical quality indicator.

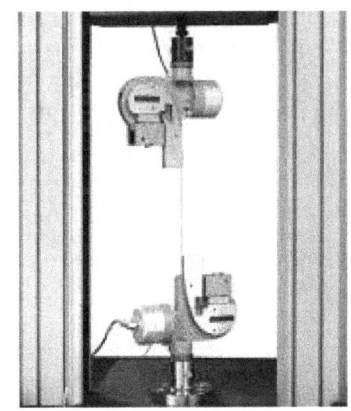

Fig.13 Tensile Strength Tester

In standard laboratory practices, a selected number of fibers are tested based on a statistical method for determining the sampling size, and the average values of related properties are reported.

Stress-Strain Curve: *Fig.14* provides an example set of data obtained from a tensile test. The value on the Y-axis represents the stress, and the value on the X-axis represents the strain, hence the name "Stress-Strain (SS) curve."

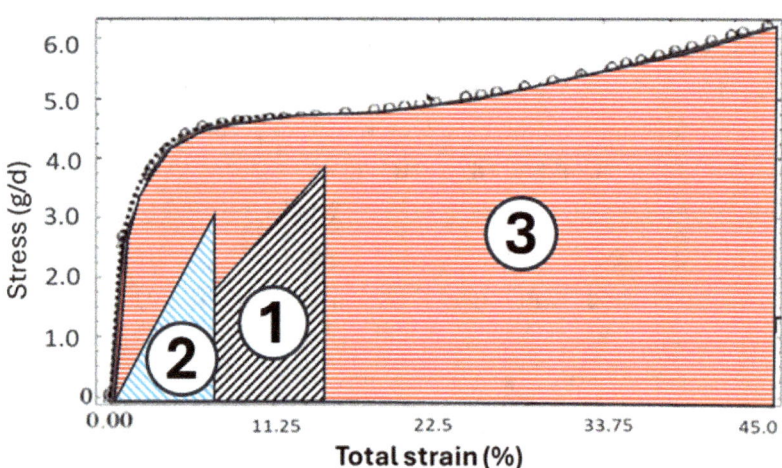

Fig.14 Breaking Energy Measurements

(1) Breaking Strength of Cotton - Wet
(2) Breaking Strength of Cotton - Dry
(3) Breaking Strength of Polyester

Fiber	Tenacity (g/denier)		Elongation (%)	
	Dry	Wet	Dry	Wet
Cotton	2.4 ~ 2.9	3.1 ~ 3.6	7 ~ 9	12 ~ 14
Polyester	4.8 ~ 6.0	4.8 ~ 6.0	44 ~ 45	44 ~ 45

The SS curve is highly useful for a wide range of analyses on fiber characteristics as it offers comprehensive insights into the physical behaviors and internal structures of the tested materials. Given the scope of this book, I will not delve deeply into the technical analysis on this particular set of data. Instead, I will focus on analyzing the properties related to the dust creation between polyester and cotton.

Analysis: The SS curves for the cotton and the polyester reveal distinctly different patterns. The cotton curves typically form triangles, while the polyester exhibits curved shapes with multiple inclinations before breaking. The shape of the polyester on *Fig.14* reflects its internal polymer structure characteristics, particularly its degree of crystallinity, a trait shared with other synthetic fibers: the initial steep slope of the polyester indicates the strength required until the tensile force reaches a yield point, the point at which the crystallinity no longer provides the dimensional strength to maintain the original fiber structure. Beyond this point, the curve's slope decreases, showing that the fiber stretches more easily with the applied unit force. The degree of the stretchiness is the function of the non-crystalline parts of the polymer structure. Together, these characteristics contribute to the total energy required to break the fiber.

In contrast, the cotton exhibits a straight line until it breaks, a typical behavior for most plant-based fibers. This can be largely attributed to the cellulosic structure of cotton, which has a more uniformly distributed structure throughout and lacks the combination of the distinct crystalline and non-crystalline regions found in synthetic fibers.

Industry Specifics (1) – End

The critical point for the readers to understand in this analysis is that the area under each graph represents the breaking energy of the tested fiber. Higher breaking energy indicates greater resistance to physical forces, making the fiber less prone to breaking (or creating dust) during manufacturing and consumption.

Variability: Prior to delving further into the analysis, it is imperative to acknowledge an inherent variability in this type of data. The characteristics of natural fibers such as cotton can exhibit significant fluctuations contingent upon various factors, including geographical origins and climatic conditions during cultivation, specific genetic variants of cotton plants, farming methods, etc. Similarly, synthetic fibers like polyester demonstrate considerable variability in their properties, largely influenced by diverse manufacturing parameters.

Consequently, various scientific literatures and other sources of information present a wide spectrum of data pertaining to the physical characteristics of these fibers. This variability underscores the importance of cautious interpretation and contextualization of the data of this nature.

Notwithstanding these variations, the dataset selected and used in *Fig.14* represents the values typically accepted within the industry.

Strength Comparison: Notably, the polyester maintains consistent stress and strain characteristics whether dry or wet, as illustrated in the table of *Fig.14*. This is due to the hydrophobic nature of polyester and its strength is unaffected by moisture as a result. In contrast, the cotton exhibits significant differences between wet and dry because of its strong affinity for water, attributed to the presence of the hydroxyl groups in cellulose and the absorption of water significantly increases the cotton's strength.

The breaking energy of the polyester, represented by the red area on **Fig.14**, is significantly higher by a factor of 5 to 10 times than that of the cotton when wet, represented by the area under the graph in black, and when dry, by the graph in blue respectively.

In most situations other than during laundry, cotton remains relatively dry in consumption, therefore, the strength difference between the two fibers in real-world usage, which relates more closely to the creation of the dust in the air, would be closer to 10 times than 5.

This significant strength disparity can be understood with a simple analogy: **if a polyester shirt produces 1 gram of microplastics, an equivalent cotton shirt could generate nearly 10 grams of dust.**

As mentioned earlier, variations in these test results can arise from numerous factors. However, most factors favor polyester in the aspect of the dust creation because its manufacturing variables can be easily controlled and adjusted to create more favourable outcomes, whereas cotton is subject to many uncontrollable factors like weather conditions, specific species, etc. This naturally leads to the discussion of the next subject, where much of the variables are accounted for and result in stark differences of the dust quantities in real-world circumstances between natural and synthetic fibers.

Industry Specifics (2) – Start

Professional Experiences on Dust

During my tenure as R&D engineer in South Korea, I had an optimal opportunity to gain a profound understanding of the dust creation from different fiber types.

Quality Issues related to Dust: As a yarn manufacturer, the company I worked for occasionally received quality complaints from its customers of the weaving and knitting factories. One of the most significant and costly issues of all complaints was an excessive dust creation by some of the company's yarn products during the manufacturing processes of its customers.

Under normal circumstances, it is natural to observe some dust as yarns pass through the hard surfaces of machinery parts, often made with metals, ceramics, plastics, etc. However, if the fibers comprising the yarns are weaker than others, they produce more dust. Excessive dust becomes particularly problematic for weavers and knitters, as it accumulates and obstructs the intended passages of the yarns during their manufacturing processes. This obstruction can lead to the breakage of both yarns and small machinery parts such as needles, causing manufacturing stoppages and resulting in a material loss and waste of labor hours.

Weavers and knitters may tolerate a certain level of operational inefficiency, but they will complain when it exceeds an acceptable level. The financial and reputational damages from such complaints prompted the company to mandate me with comprehensive studies to understand the causes of wide variations in the dust quantities in some manufacturing batches of yarns, which subsequently led to customer complaints.

Yarn Blends and Dust Quantity Analysis: The company produced yarns with a wide range of fiber compositions, including 100% Cotton, 100% Wool, 100% Polyester, 100% Acrylic, as well as various blended fibers. The blend compositions varied significantly depending on the desired properties of finished textiles, ranging from blends like 95% Wool and 5% Polyester to the reverse, and everything in between as well as widely different other fiber types mixed together. As a result, the company offered hundreds of yarn assortments to its customers.

To better understand the variability in the dust generation across this wide spectrum of products, I developed a strategy to measure the nominal dust quantity for the yarns made from homogeneous materials (such as 100% Cotton, 100% Wool, 100% Polyester, etc.) and used the results as the baselines of the dust creation characteristic of each fiber type in normal quality. Then, I tried to correlate the baselines to the variation when it was blended with other fibers in varying proportions. This approach allowed me to understand how each fiber type contributed to the dust creation, regardless of specific blend ratios.

Invention of Dust Measurement Test Device: One of the challenges I had to overcome on tackling this task was accurately measuring the quantity of the dust in a controlled environment. In the open air of a typical manufacturing facility, it was impossible to obtain precise measurements. For example, enclosing a large weaving machine in an airtight bag to collect the dust over a certain period while the machine was running would have been both impractical and extremely costly. Additionally, this method would have resulted in significant dust loss when opening and closing the bag.

To address this, I developed a compact test device that fits comfortably within the company's laboratory, as illustrated in *Fig.15*. This device featured a series of needles arranged to simulate the tensile and frictional forces subject to the yarns in conventional knitting machines. When a yarn specimen entered the chamber, it was subjected to a preset degree of physical forces before being wound on the other side. The adjustments for creating different levels of physical forces were made through the speed adjustment of the winding spool and different passage arrangements within the chamber.

Fig.15 Test Equipment for Yarn Dust Collection

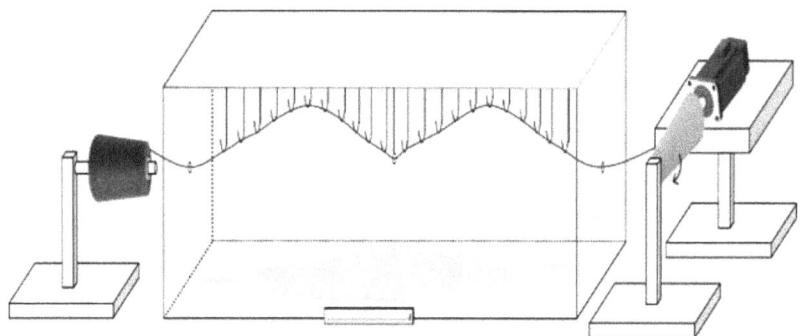

In 1993, I was granted my first patent in South Korea for the device. With it, I established a nominal range of the dust quantity for varying yarn types based on different manufacturing parameters such as fiber types and blend compositions, yarn thicknesses and twist variables, among others. If a batch of yarn produced by the company exhibited a dust generation characteristic considerably exceeding the nominal range, the batch was segregated and destroyed. This approach improved the company's operations by minimizing the customer complaints of the nature.

Industry Specifics (2) – End

Generalization of Dust Quantity: Through this experience, I established a strong correlation between the breaking energies of fibers and the amount of dust generated during subsequent manufacturing processes. For example, under identical yarn-making conditions and produced in the same pilot spinning machines, a cotton yarn created multiple times higher quantities of dust than a polyester yarn, assimilating the data presented in *Fig.14*. This observation was applicable for other fiber types, and it was clear that the weaker strength of natural fibers was translated directly to the higher quantity of dust created. My earlier assertion, *"if a polyester shirt produces 1 gram of microplastics, an equivalent cotton shirt could generate nearly 10 grams of dust"*, was in part based on the conclusion obtained from this study, exemplifying the nature of scientific studies proportionally reflected in real-life situations.

Section 4   Unfortunate Disposal

While many people may not give much thought to the environmental impacts of disposing of textile goods, there are numerous critical factors to consider. These range from the implications of the residual chemicals contained in the waste affecting our environment to humanitarian crises caused by the export of textile waste from affluent countries to underdeveloped and developing countries. Some of these disposal practices may be shocking to those who are unaware.

4.1   Donation Myth

Many people assume that the textile goods they drop off in community recycling boxes or donate to charity groups are reused by those in need. However, the reality is starkly different from this common perception. In a 2018 episode of "Marketplace", aired by the Canadian Broadcasting Corporation (CBC), it revealed that only 1% of the textile waste is reclaimed and recycled in a "real sense" of recycling. The fate of a large portion of it is mind-boggling and even exposes a troubling side of humanity.

Unintended Destinations: Strict environmental regulations in affluent countries make it difficult and expensive to locally incinerate or bury large quantities of textile waste. Consequently, these textiles are shipped to underdeveloped and developing countries far away from the origins of the waste. While a small fraction of them may be reused by the locals in the destinations, the majority are burned or buried without necessary environmental safeguards, such as monitoring for volatile organic compounds (VOCs) or proper analysis and control on soil and water contamination.

It is a troubling irony that thick sweaters and winter coats from wealthy northern countries end up in nations near the equator with hot weather, reflecting a typical "Not In My Back Yard" (NIMBY) attitude of the citizens of wealthy countries.

Misperceptions and Environmental Impacts: As the environmental awareness grows across the globe, people are becoming more mindful of their ecological footprints. For example, many would consider it unsustainable to toss a plastic bottle into regular garbage. Yet, these same individuals may dispose of their unwanted clothing in recycling bins or donate them to charity without the same level of concern as they are unaware of the resulting reality.

This misperception carries significant environmental consequences. Beyond the carbon footprint from transporting the waste across oceans, the textiles contain residual chemicals that persist in nature, causing long-term harm, a topic explored in detail in **Chapter III: Lifecycle Assessment - Disposal** section.

Humanitarian Crisis: Perhaps the most distressing aspect of this issue is the social injustice it perpetuates. People in affluent countries often ignore the harsh living conditions of those in the nations that receive their waste. In these regions, mountains of textile waste pile up along riverbanks or are burned in areas like school playgrounds.

I am convinced that if people in the origin countries of the waste were fully aware of the impacts from these practices, they would not accept such senseless act of the environmental and social harm.

4.2 Decomposition Myth

Many environmental interest groups often target synthetic fibers, due to their prolonged persistence in nature. It is commonly understood that natural fibers decompose and return to the environment much faster than synthetic fibers. However, while this basic scientific fact is true, there are many other critical factors that must be considered to fully understand the global environmental impacts of natural versus synthetic fibers.

Residual Chemicals: Although natural fibers like cotton (cellulose-based) and wool (protein-based) degrade more quickly in nature, the chemicals used during their manufacturing and those left in the textile waste present a significant challenge in the environmental management. For instance, various colors in the textile products indicate the presence of the dye pigments and numerous other processing chemicals. These substances can leach into soil, contaminating runoff and groundwater in landfills, thereby entering the ecosystem. When incinerated, they pollute air and can spread globally through the atmospheric circulation.

While more detailed information will be provided in **Chapter II: Basics** and **Chapter III: Lifecycle Assessment**, it is important to recognize that natural fibers often require significantly larger amounts of harmful chemicals during production compared to synthetic fibers, primarily due to their heightened chemical reactivity. Furthermore, the chemicals used for natural fibers are typically more reactive, as they are formulated to interact with the chemically active functional groups in these fibers, whereas the chemicals used for synthetic fibers generally do not depend on chemical reactions in the same manner.

Thus, understanding the quantity and toxicity of the chemicals involved in both natural and synthetic fibers is crucial for accurately assessing their true environmental impacts.

Environmental Impacts

Toxicity of Pesticides for Farming Natural Fibers: Natural fibers, particularly the plant-based ones like cotton, are exposed to toxic chemicals from the very beginning of their lifecycles. Pesticides and fertilizers used in farming are designed to eliminate living organisms like insects and bugs and are therefore highly toxic. In high concentrations, they can even significantly harm humans and large animals. In fact, the toxicity of these chemicals far exceeds than that used in most other textile manufacturing processes such as dyeing. Much of this toxicity remains in the environment or in the textile materials themselves for periods that can far exceed multiples of human lifetime.

Dye Types Overview: Among all the stages of textile production outside the cultivation of natural fibers, the dyeing is typically most chemically intensive and creates the highest environmental impacts. Different types of fibers require different dyes and associated chemicals based on their chemical compositions. For example, cotton and wool require highly reactive dyes, known as "Reactive Dyes", while polyester uses a different type, known as "Disperse Dyes". As the name indicates, the reactive dyes contain chemical elements and substances which promote reactions with the chemical groups of the targeted fibers whereas the disperse dyes are designed to physically impregnate into the polymer structures of synthetic fibers.

Readers familiar with hair dyeing and washing plastic food containers can easily draw parallels to textile dyeing. Like wool dyes, which are reactive and designed to chemically bond with the protein in wool, the hair dyes also involve a chemical reaction with the protein contained in the human hair. The dyed hair color fades over time relatively quickly, and proportionally with the frequency of washing. This can be compared with the colors on plastics, often found in food container lids, etc. These colors never fade regardless of how many times they are washed because they do not interact with water and does not shed color pigments. More detailed information on the dyeing related aspects will be provided in the following discussions including the dedicated **Chapter II: 10.4.2 Dye Types and Toxicities.**

Chemical Reactivity and Environmental Impacts: In general, the environmental impacts of the reactive chemicals used in textile production relate to both their chemical reactivity and their persistence in the environment, either in their original forms or as the byproducts of associated chemical reactions. These impacts are particularly significant when more reactive substances, such as salts, heavy metals, etc., are involved in the process. On the other hand, less- or non-reactive chemicals, like the disperse dyes, tend to have environmental impacts more due to their persistence on their own rather than chemically reacting with their surroundings or creating byproducts.

Natural Existence and Densification: Although many of the chemical elements and compounds used in the textile manufacturing exist naturally in the environment, their concentrations are typically low and do not pose health risks to humans and animals under usual circumstances. However, when these are artificially concentrated to achieve specific characteristics, their toxicity increases, posing significant threats to the living organisms and environmental sustainability. Pesticides, fertilizers, dyes, and various other chemicals used in the textiles are examples of these concentrated substances. Some of these compounds are known to cause severe health effects, including cancer, coma, and seizures.

***

Different chemical treatments required for natural versus synthetic fibers, while both are harmful, can lead to significantly different environmental outcomes. This topic is explored in detail throughout **Chapter II** and **Chapter III**.

4.2.1 Duration of Decomposition

There is a myriad of information online regarding the decomposition durations of synthetic materials, but much of it lacks scientific validity, leading to widespread public misconceptions and associated environmental harm. An example is illustrated in ***Fig.16***, where it is claimed that plastic bottles and diapers last for 450 years in nature. Other sources suggest even longer durations, some stating that plastic bottles may persist for 650 years or even indefinitely. These claims, often ranking high in various search engines like Google, contribute to the misinformation crisis we face today – more discussions to be presented in **Chapter IV: Misinformation Crisis**.

Fig.16 Example of Plastic Duration in Nature on Internet

Validity of Information

Plastic was invented in 1907 and has been around for just over a century. Naturally, scientific and analytical minds might question the claims about plastic persisting in nature for hundreds of years. Some might argue, "*If the modern science can calculate the Earth's age at 4.55 billion years*, estimating plastic's longevity in the range of hundreds of years shouldn't be difficult.*" However, this line of thinking overlooks the fundamental principles behind the underlaying science.

> * The figure established by Dr. Clair Patterson and widely accepted in the scientific community

Radioactive vs. Chemically Inert Materials: The methods used to determine the age of Earth and also ancient artifacts rely on the radioactive decay of certain elements. For example, Dr. Patterson used the half-life of a lead isotope to calculate the Earth's age where the half-life refers to the time it takes for half of the atoms in a radioactive substance to decay.

However, for these calculations to work, the material in question must contain a "radioactive" element. Plastics, by contrast, are chemically inert and do not undergo radioactive decay. In fact, if plastics were radioactive, it would be too dangerous for us to use them so close in our daily lives due to high health risks. This means that the science behind calculating the Earth's age is fundamentally different from estimating the duration of the plastic decomposition. While there are scientific methods for studying the plastic degradation under specific conditions, these methods are based on vastly different scientific principles and cannot be used to accurately predict plastic's decomposition timeline in the context of several hundreds of years. While related scientific details will be discussed in subsequent discussions, it is clear that widely circulated information on plastic decomposition such as the earlier "450 years" example lacks a solid scientific basis.

Plastic Decomposition and Evidence

As such, one common misunderstanding in the textile sustainability is the perceived longevity of synthetic fibers (or plastics in general) in nature. To better understand this, let's travel back in time to Seoul, South Korea, between 1978 and 1993. During this period, Nanji-do served as a massive landfill for the greater Seoul area, receiving an average of 3,000 truckloads of household waste daily. By the time it closed, it was the tallest landfill in the world, containing the total of 93 million tons of waste.

Personal Experience: I have a personal anecdote which relates to this site of Nanji-do. In 1995, as a young professional freshly landed in Seoul, I decided to save money on driving lessons by practicing at Nanji-do. Unlicensed driving instructors used the landfill site to draw parking and practice routes on flattened areas on top of the waste and offered lessons at much cheaper rates than licensed ones in the city. I commuted there for the lessons, enduring severe odor coming from the decaying waste and observed firsthand the compositions of the waste, including large amounts of plastics (see *Fig.17* for the examples of colorful plastic waste).

Fig.17 Nanji-do in the 90's

After I immigrated to Canada in the late 1990s, Nanji-do faded from my memory. Then, years later, I came across a news article reporting that the landfill had been transformed into a part of the Seoul Sangam World Cup

Stadium Park. Interestingly, a portion of Nanji-do was designated for a power plant that generates electricity using the gas from the decomposition of the household waste, including plastics. From the landfill's closure to the extraction of this gas, it took about 30 years.

Fig.18 Nanji-do NOW

Image: Seoul WolrdCup Stadium Park

Image: Power Plant

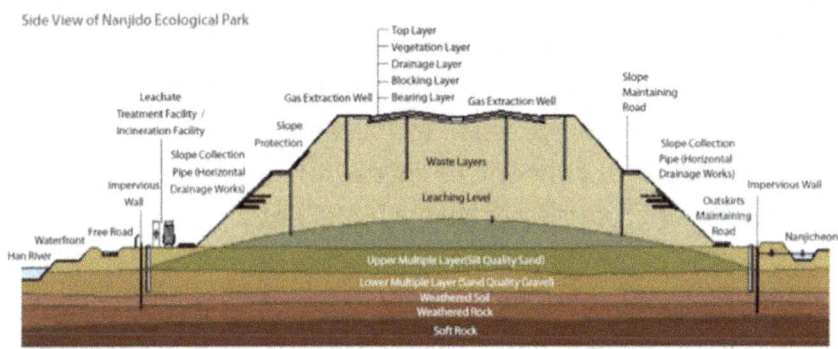

Our Own Observations: Now, let's fly out of Seoul and return to our own backyards and surroundings. Many of us have encountered plastic pieces left in the environment, perhaps picked up while walking or gardening. If you have purchased plants from a local store, you may have noticed that they often come wrapped in black plastic bags around the soil.

Significant Strength Degradation: Our experiences with unused or freshly used plastic bags give us a baseline for the typical strength of plastics in daily use. Compare this to the plastics used to wrap the soil, which often show significant strength degradation after being in contact with the soil for a relatively short time. In the perspective of the plastic manufacturing, it is implausible to intentionally make the initial quality of the plastics so fragile due to their inherent characteristics related to the polymerization principles. Although not a perfect "apples-to-apples" comparison, we can easily estimate that these plastics weakened over time as they tear or puncture far more easily than the baseline reference based on our experiences, suggesting that the decrease in the strength was associated with the soil contact.

Science of Degradation: In scientific terms, the degradation of the plastic strength indicates that the polymer chains inside the plastics broke down. To illustrate this concept, imagine a fishnet: if a few of its connecting parts are broken, the net's overall structure becomes compromised and the strength decreases. Similarly, when the bonds within the plastic's polymer chains break, the material loses its strength.

Practical Experiment: A curious mind can easily verify this phenomenon at home: Bury a piece of plastic bag under wet soil in a shaded area and leave it for a month or two while keeping another identical piece in a kitchen cabinet. After a certain duration, the buried plastic will have significantly lost its strength, while the stored plastic will remain intact. This experiment is a practical way to observe the plastic degradation without having to wait "hundreds of years", the durations often cited on the internet - I recommended using plastic bags used in grocery shopping for this experiment because the effects will be more readily noticeable in a relatively short time frame and can be verified without the need of laboratory equipment.

***

As mentioned earlier, plastics are not radioactive and chemically inert. Thus, the plastic degradation is based on a different scientific principle than the radioactive decay method of accurately calculating the ages of ancient artifacts. It is in fact due to the existence of the plastic-eating bacteria in nature. In certain conditions, such as chemically active landfills, these bacteria thrive in higher densities and exhibit increased level of activities. Certain chemical compositions, temperatures, and pressures present in such landfills significantly aid these bacteria in degrading plastics at a much faster rate than what many sources suggest on the internet. The Nanji-do landfill is a prime example of this accelerated degradation process. Furthermore, the subsequent discussion of creating the artificial conditions of plastic decomposition is another example.

Artificial Conditions of Decomposition: Another piece of evidence comes from various "biodegradability" claims made by many companies today on certain plastics. In some laboratory or manufacturing environments, specific conditions such as particular bacterial compositions, temperature, pressure, etc., can be created for plastic degradation, the conditions reflective of actual landfills (anaerobic degradation), or left in the open air (aerobic degradation).

While humans can create and adjust these conditions for accelerating or decelerating the degradation rates, they do not create the bacteria themselves. These organisms naturally exist in the environment and constantly at work to break down the materials left in nature. By definition, therefore, all plastics are biodegradable and many biodegradability claims, unfortunately, are simply the exploitation of the public interests on the environmental sustainability with the misrepresentation of the underlying science for their benefits.

Variabilities of Plastic Decomposition: Different pieces of plastics degrade at varying speeds. The form factor of plastics whether they are thin and light like plastic bags and bottles or thicker and heavier like food containers affects how easily the bacteria can access and break down the polymer chains. Additionally, the density of plastics plays a crucial role: the higher the density, the more challenging it is for the bacteria to penetrate the polymer structures and break the links.

4.3 Biodegradability

The term "biodegradability" has recently become a major source of misinformation and greenwashing by many organizations, exploiting the public's keen interest and pursuit of the environmental sustainability. In this subsection, I will debunk some myths surrounding "biodegradable" plastics.

Before diving in, it is important for the public to understand that the term "biodegradable" by itself does not necessarily indicate an environmental benefit. There are many factors involved in this simple term, and without understanding these nuances, the public can be misled as the term is often used out of context.

Definitions

The first step is to clearly define the "biodegradability". A plastic is considered biodegradable if it breaks down under natural conditions. As discussed earlier, however, all plastics are technically biodegradable due to the activities of the plastic-eating bacteria in nature. Given enough time, all plastics will eventually degrade. Therefore, this discussion begins with highlighting the inherent flaw in "biodegradable" claims.

Context: In the context of labeling a plastic as "biodegradable", specific conditions, typically in laboratory settings, are used to assess its degradation characteristics in a certain form factor under controlled environment. These conditions involve carefully regulated factors such as bacterial activity, temperature, pressure, and chemical composition in relation with a time factor.

However, this approach has limitations. For example, a thin plastic might be classified as "biodegradable" based on the test results obtained under certain conditions, while a thicker piece of the same polymer may not be, due to the differences in how easily bacteria can access the polymer chains and the rate of degradation. Additionally, real-world conditions in landfills vary significantly across the globe, making it unrealistic to expect that a plastic labeled "biodegradable" will degrade the same way in diverse environmental conditions as it did in the controlled test environment.

Thus, it is important to approach the term "biodegradable" with caution, as it can be easily misunderstood or misused by profit-driven organizations. The public should also be aware that they should not mistakenly assume plastics without a "biodegradable" label do not naturally degrade.

In this discussion, I use the term "biodegradable" reluctantly and only within the context of common understanding in the industry, while acknowledging the limitations of its definition.

Bioplastic: The term "bioplastic" refers to the plastics that are produced from natural biomass sources of either plants or animals such as corn starch, woodchips, proteins, etc., or the presence of bacteria in their synthesis. However, not all bioplastics can be identified as "biodegradable" according to the definition mentioned earlier. For example, bio-based PET (polyethylene terephthalate) is a non-"biodegradable" bioplastic. Even though it is synthesized with the presence of bacteria, the chemical structure of the bio-based PET remains identical to its fossil fuel-derived counterparts, including its decomposition properties. Some studies show that certain bioplastics can be manufactured with a lower carbon footprint than their fossil counterparts while other bioplastics' processes are less efficient and result in a higher carbon footprint than fossil plastics.

Although more comprehensive understanding is required rather than simply accepting the results of these studies, it can be easily concluded that we cannot automatically assume an environmental benefit from the term of "bioplastic", the theme which applies for the remainder of the discussions of this subject.

Oxo-Degradable Plastics: Another example is "oxo-degradable" plastics. Despite their seemingly eco-friendly name, they are essentially conventional plastics with additives known as pro-degradants that accelerate oxidation process under specific conditions involving oxygen. While oxo-degradable plastics break down more quickly when exposed to sunlight and oxygen, conditions typically found in open air rather than in landfills or water, this process merely fragments the plastic into microplastics. As a result, the perceived benefit can lead to other environmental and health risks as microplastics contaminate air and water, ultimately entering the food chain and posing threats to both human and animal health.

Compostable Plastics: In some cases, compostable plastics are labeled as "biodegradable". Compostable plastics decompose within a specified timeframe under controlled conditions typically found in industrial composting facilities. These facilities create optimal environments for decomposition, including high temperatures, controlled pressure, and specific chemical conditions. Accelerating the decomposition rate of otherwise-naturally-degrading plastics renders this factory-based decomposition process economically inviable. Thus, there are very few such facilities, and they are generally used for specialized purposes.

Degradability of Plastics: An important aspect of these discussions is the natural degradability of all plastics, as repeatedly emphasized throughout this section. To substantiate claims of more favorable environmental impacts, a "biodegradable" plastic must be proven to degrade in natural conditions, such as in landfills, air, or water, considerably faster than other plastics, and without creating harmful side effects.

With this premise in mind, it is beneficial to review the scopes of a typical test method used to label plastics as "biodegradable" and other related claims.

Industry Specifics (3) – Start

ASTM D5511 – 18

Excerpts of the test method includes the following statements:

> *This test method is designed to yield a percentage of conversion of carbon in the sample to carbon in the gaseous form under conditions found in high-solids anaerobic digesters, treating municipal solid waste. This test method may also resemble some conditions in biologically active landfills where the gas generated is recovered and biogas production is actively promoted by inoculation (for example, code position of anaerobic sewage sludge, anaerobic leachate recirculation), moisture control (for example, leachate recirculation), and temperature control (for example, short-term injection of oxygen, heating of recirculated leachate).*
>
> **Claims of performance shall be limited to the numerical result obtained in the test and not be used for unqualified "biodegradable" claims.** *Reports shall clearly state the percentage of net gaseous carbon generation for both the test and reference samples at the completion of the test. Furthermore, results shall not be extrapolated past the actual duration of the test.*

Industry Specifics (3) – End

As a result, the outcomes of this test method **can only be presented as numeric data and cannot be used to justify "biodegradable" claims.** Several other test methods under similar scope exist globally.

Corporate Misconduct: Unfortunately, many companies exploit the fact that consumers are unlikely to fully understand the implications of such claims, taking advantage of this loophole. While this approach is clearly unethical, such corporate misconduct is not uncommon. Companies often present limited information without proper context, leading consumers to instant misperceptions. This behavior is typical of the organizations that prioritize short-term gains over transparency, due diligence and honest effort to truly benefit the environment.

Even if we extend the benefit of doubt and assume that these companies lacked the necessary knowledge and acted unknowingly, this does not absolve them of their social and environmental responsibilities. A couple of hours of thorough reading would be enough to grasp the basics of the subject matters and guide them to avoid being engaged in such acts. Failing to put in even this minimal effort and commitment can only be described as misconduct and greenwashing. Having spent decades in the global textile industry, I find it deplorable to witness this level of dishonesty even from some of the most reputable brands which undeservedly enjoy a high level of consumer trust.

Poor Business Practices and Greenwashing

Companies making "biodegradable" claims must present all relevant information clearly and transparently, ensuring consumers can easily understand the scope and limitations of the tests utilised, rather than simply labeling products as "biodegradable". Several countries, including the USA, European Union, Australia, and Japan, have regulations to prevent companies from using "biodegradability" claims as a form of greenwashing. However, many claims are still made without proper substantiation or adequate context, leaving consumers misinformed.

"Biodegradable" Thermal Insulation?: Some companies exploit other loopholes in "biodegradability". For example, a company in the thermal insulation industry claims that one of its products is "biodegradable".

In manufacturing, thermal insulation materials are sandwiched between lining fabrics and enclosed within the structure of finished goods. This means that these materials are not typically exposed to the external environment in nature. The lining fabrics themselves can vary widely in characteristics - some are thin and lightweight, while others are thick and heavy, often coated with polymeric membranes, all acting as barriers against bacterial access.

When testing "biodegradable" materials, they are evaluated in isolation, without considering other components of finished products. Therefore, this approach of using the "biodegradable" conclusion obtained from such method works for single-layer items, such as shirts or pants, where the fabrics in question can be directly exposed to similar decomposition conditions in nature. However, for multi-layered structures like winter jackets or heavy blankets that contain thermal insulation materials inside, any "biodegradability" claim about the insulation material must also account for various lining fabrics used.

Given the wide variability in these fabrics and structures of finished products, I can confidently assert that verifying the "biodegradability" of thermal insulation materials in nature is virtually impossible.

Just Another Exploitation Example: Yet, the thermal insulation company continues to make their unsubstantiated "biodegradability" claims without addressing these concerns. As previously discussed, the public's lack of knowledge enables the prevalence of numerous cases of exploitation by these greenwashing organizations.

***

As a consumer, I do not consider accelerated biodegradability of plastics to be beneficial to my lifestyle. If a plastic's biodegradability is significantly enhanced compared to regular plastics, I would be concerned about its degradation during use. I keep most of my textile goods for decades, such as the very first suit I wore for my university graduations, still hanging in my closet and occasionally worn. I would not want them to degrade in my closet and continuous use.

Section 5   Unfortunate Recycling

Many people form opinions and perspectives on various aspects of life based on single-word descriptions or short phrases. "Recycling" is one such example. Given the global importance of the environmental sustainability, one might expect questions like, "*Is recycling done without creating negative environmental impacts?*" to be common. However, this is often not the case.

Face value: Even among the most educated members of our society, there is a tendency to accept convenient notions and purchase textiles with such brief sustainability claims. Labels such as "*Recycled*" or "*Made from x number of plastic bottles*" are often taken at face value.

Serious Gap: There are several key reasons for this phenomenon. One is that consumers often place too much trust in the companies making various sustainability claims. Furthermore, consumers may not be trained to ask critical questions and often lack diligence to seek truth on more complex issues which do not create more immediate and observable impacts in their lives. Unfortunately, environmental matters are of such characteristics. Together, these factors create a significant gap that allows for numerous corporate misconducts, where untrained consumers are exploited.

Given the prevalence of greenwashing, even by some of the most trusted organizations globally, the public's overly simplistic approach to various sustainability claims does little to benefit the environment. It is crucial for consumers to take a more critical stance and demand transparent, comprehensive information in order to make informed decisions that truly benefit the environment.

Recycling Myth

Plastic recycling can be conducted in various ways, but the most conventional methods unfortunately result in critically negative environmental impacts. In fact, we do not even know if there is a clear net benefit from the current global recycling efforts. For many readers, this statement may be surprising, as most consumers believe that purchasing goods made from recycled contents is beneficial for the environment. Here, I present the information that may challenge this common understanding.

Importation Ban on Plastic Waste: Many people may recall the garbage crisis that most Western countries experienced in the late 2010s, which was precipitated by China's total ban on the importation of plastic waste, the primary source for plastic recycling - For years, advanced economies had been exporting large quantities of their plastic waste to China and other countries, and this sudden ban led to local waste crises.

China's Gradual Actions and Similar Actions by Other Countries: Although the ban seemed abrupt to the Western world, the Chinese government had initiated related measures as early as 2013 with the CHINA GREEN FENCE policy. This was followed by the CHINA NATIONAL SWORD policy in 2017, culminating in the total ban in 2018. Subsequently, many other countries, such as Vietnam, Malaysia, Thailand, etc., implemented similar measures, focusing on processing only their own domestic plastic waste. The actions taken by these governments stem from a clear understanding of the negative environmental impacts associated with the conventional recycling processes and are now spread to a large number of other underdeveloped and developing countries around the world.

In this discussion, I will elucidate four main areas of negative impacts caused by the conventional methods, accompanied by the flow map depicted in *Fig.19*.

Fig.19 Overview of Conventional Plastic Recycling

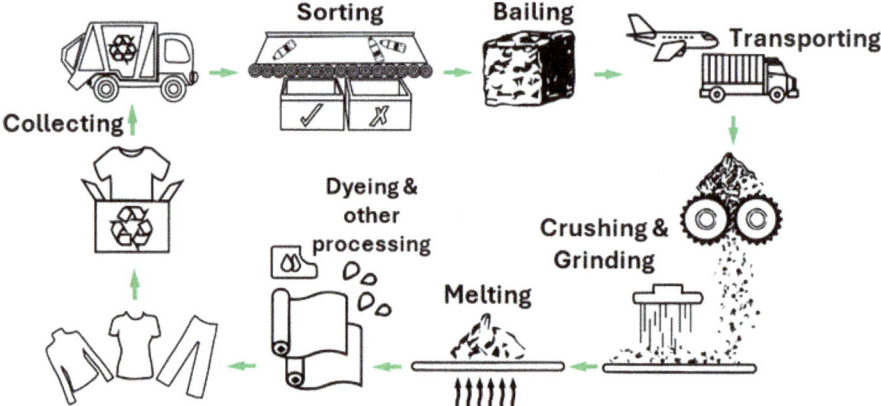

Inefficient Carbon Footprint: The processes result in a significantly inefficient carbon footprint due to the transportation of plastic waste across international oceans, as most recycling facilities are located in countries far from the consumption centers where the waste is generated. Given that a substantial portion of the plastics used in the products and packaging originates from these distant countries to begin with, shipping the waste back to the regions of origin exacerbates the inefficiency in the carbon footprint management

Microplastics in Manufacturing: Most conventional recycling is carried out through a process known as "crushing" and "grinding", shown by "2" in *Fig.19*. This method involves applying substantial physical force to break the waste into small pieces, which are then submerged in a large pool of water.

The purpose of this process is to remove dirt and foreign materials contained in the waste. Observations at the storage yards of recycling facilities readily reveal such heavy contamination present in the plastic waste, waiting to be processed.

Fig.20 Contaminated Plastic Bottles

Given the diverse shapes, sizes, and types of contamination present, the only commercially viable way is to crush the plastics and break them into small pieces, then use water to wash away the contaminants. *Fig.21* illustrates the stage of the process where the crushed plastics are deposited into a large quantity of water.

Fig.21 Crushed Plastics

Additional details of such process can be found in the YouTube video link below.

https://www.youtube.com/watch?v=URfgzIa1UfY&t=35s

The application of heavy physical forces to break down the plastics inevitably generates a significant amount of debris, including microplastics.

Strength Factor and Environmental Harm: Despite the extensive cleaning efforts, which involve heavy water and electricity usage, microplastic generation, and potential chemical use to dissolve certain types of contaminants, it is virtually impossible to remove all foreign materials. As a result, significant quantities of these contaminants persist throughout the subsequent recycling processes and become integrated into the recycled fibers.

These contaminated recycled plastics exhibit distinctive colors. Higher proportions of contaminants result in a more greyish tone in the polymer pellets, as shown in *Fig.22*, which are then converted into fibers, carrying the grey tone over. This color difference cannot be detected by consumers, as various efforts are made during subsequent dyeing processes to conceal such imperfections.

Fig.22 Recycled Pellets (left) vs. Virgin Pellets (right)

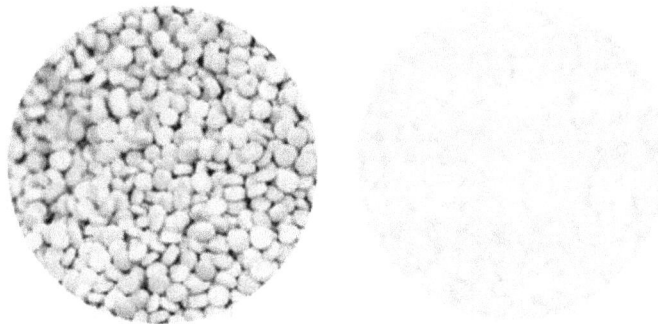

The presence of contaminants in the recycled fibers adversely affects their strength as it hinders the formation of polymer structures during the subsequent melting process. As a result, most recycled fibers are invariably much weaker than their virgin counterparts. In some cases, the tensile stress and strain, and consequently the total breaking energy, of recycled fibers are less than 50% of those of virgin fibers.

The reduced strength of recycled fibers contributes to shedding more microplastics and chemicals during manufacturing and consumption as textile materials encounter various physical forces during their lifecycles.

Harmful Chemicals: Recycled fibers undergo identical manufacturing and consumption processes as virgin fibers, including chemically intensive dyeing process. It is well-known within the industry that different dyeing formulations must be used for recycled fibers to compensate the varying degrees of contamination to achieve desired color tones and hues whereas pure virgin fibers do not present such issue. This necessity may lead to increased usage of color pigments and other chemicals to counteract their inherent greyish tones.

Moreover, recycled fibers may lose color pigments and other chemicals more readily during use as the contaminants can impede the fixation of these chemicals within their polymer structures, resulting in a higher quantity of chemical release into the environment.

***

In conclusion, there is no clear evidence to suggest that the conventional recycling efforts yield a net benefit in the overall environmental management. In my view, the disadvantages can easily outweigh the benefits. Increased quantities of microplastics alone could justify ceasing these activities altogether.

I have long argued that, given today's inefficiencies in plastic recycling, resulting in considerable harm, the best option is to place plastics in landfills. Unlike textile waste, plastics generally do not contain harmful dyes or chemicals. Even colorized plastics likely contain non-reactive dyes that do not interact with the environment. While plastics may remain in landfills for extended periods, they are less likely to contribute to other significant environmental harms.

The good news is that we do have a viable solution based on the principles of successful plastic recycling. This topic will be further explored in **Chapter V: Solutions**, specifically in the **Clean Recycling Initiative™** section.

Chapter I: Post-Chapter Commentary

In this chapter, I discussed various topics related to consumers and the textile products they consume, along with the environmental issues involved. I made a concerted effort to present the relevant science in simple terms to ensure all readers, regardless of their prior knowledge or experience, can transition smoothly into gaining a more fundamental understanding of textile materials and science in **Chapter II: Basics**.

Along with many examples shared in this chapter, readers will gain deeper insights in the next chapter into the widespread misunderstandings and misconceptions that arise from oversimplified ideas, such as the flawed concept of "fast fashion" and the unsubstantiated claims that plastic bottles take 450 years to decompose. Unfortunately, these misconceptions dominate public perception and shape society's collective mindset about what is good or bad for the environment, often inadvertently undermining global sustainability efforts.

Consumers, Be Aware and Beware!: If we continue to allow these mistakes and mismanagement under the guise of environmental sustainability, we may reach a point of no return. In my view, spreading accurate, science-based information is the first step to take to mitigate the negative impacts from such undesirable actions.

Chapter II
**
Basics

Pre-Chapter Commentary

Quote of the chapter

"I reiterate that I do not deny the negative impacts of synthetic fibers. However, we must acknowledge that we do not live in a utopian world where all problems vanish by replacing one problematic material with another. My point throughout the book is that we must consider and evaluate the true overall impacts with comprehensive views and throughout the entire lifecycles of the materials in question. Favoring one material type over another based on limited scopes may result in creating more detrimental effects on the environment."

***

Building on the consumer-related topics discussed in the previous chapter, many readers may seek a deeper understanding, particularly on issues that challenge common public perceptions as I recognize that the discrepancies between the information presented in this book and widely held beliefs may provoke curiosity for some readers and strong resistance for others.

This chapter is intended to provide those with the essential knowledge needed to assess the environmental impacts of various textile materials based on accurate scientific information, rather than relying on widespread misinformation or overly simplistic notions. The contents of this chapter will cover the fundamental characteristics of different materials, their chemical compositions, implicated chemicals and manufacturing processes, with a specific focus on the environmental impacts of each.

Section 6 Overview - Fibers

In this section, we will explore the fundamental concepts of fibers, grouped into three broad categories: natural, synthetic, and semi-natural. This overview will cover the origins of these fibers, how they are produced, and their relevance to environmental sustainability. While some readers may already be familiar with certain aspects discussed here, this section aims to build on foundational knowledge rooted in relevant scientific principles, laying the groundwork for more science-based discussions in the subsequent sections and chapters of the book.

Basic Scope

Fibers constitute the fundamental unit and primary component of the majority of textile goods. Typically, fibers are transformed into yarns, then to various forms of fabrics such as woven, knitted, and non-woven materials before becoming the finished goods we consume.

Fig.23 Textile Manufacturing Flow Chart from Fibers to Apparel*

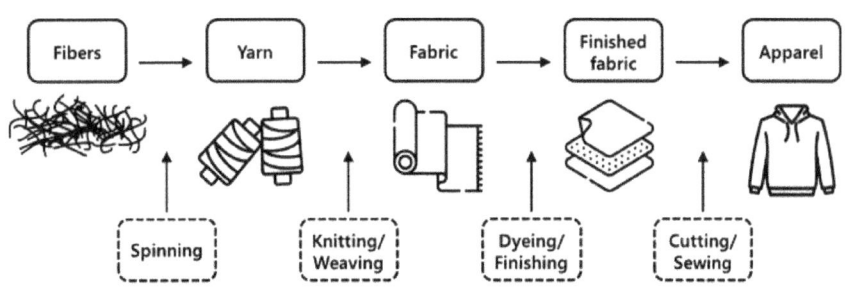

* Typical process and may vary depending on the needs of finished goods

There are notable exceptions to this generalization, including materials like leather, fur (a subset of leather), mycotextiles derived from fungi (also known as plant-based leather), which naturally form sheet structures from the growth of animals or fungi. Another exception is down-feathers (also known simply as "down"), sourced from different bird species and commonly used as filling material in winter garments, home textiles, sleeping bags, etc. These materials are utilized in their harvested form without having to go through a conversion into other intermediary materials such as yarns or processed forms of fabrics.

Transformation: The thin and elongated structure of fibers allows for their alignment and multiplication in the desired length directions. When a large number of fibers are aligned and twisted together, they form a strong, flexible, and pliable yarn. These versatile properties of yarns are transferred to subsequent intermediary materials such as woven and knitted fabrics and finished goods, enabling them to withstand physical and chemical forces throughout their lifecycles.

Fiber Types: As mentioned earlier, there are three primary fiber types: natural, synthetic, and semi-synthetic (also known as semi-natural) fibers. Natural fibers are derived from the growth of living organisms, while synthetic fibers are produced through the synthesis of petroleum-based raw materials. Semi-synthetic fibers, as the name suggests, occupy a middle ground between these two categories.

Basics Overview of Natural Fibers: The characteristics of natural fibers, such as thickness, length, strength, chemical resistance, etc., are determined by the species of specific sources, such as plants or animals. Typically, the properties of fibers within a particular species are influenced by various farming factors, including weather conditions, rainfall quantities, food sources, farming techniques, etc., prior to harvesting. Human-mediated genetic modifications may be utilized for altering these characteristics beyond their natural ranges or for benefiting farming outcomes. Such examples include genetically modified cotton, known as Bt cotton, or bioengineered silk, etc.

Basics Overview of Synthetic Fibers: In contrast, the characteristics of synthetic fibers are determined by specific polymer types and can be adjusted to reflect the desired properties of finished goods. For instance, the length of synthetic fibers can be easily adjusted with simple adjustments made during manufacturing processes; continuous fibers are known as "filaments", while those cut into shorter lengths to mimic natural fibers are called "staple" fibers. The thickness can also be as easily adjusted ranging from extremely thin strands, measured in micrometers or even nanometers, to much thicker strands such as the filaments of fishing rods and ropes for industrial use. The cross-sectional shapes of synthetic fibers can be modified from the conventional round to hexagonal, oval, hollow, or peanut shapes, among others, with equally simple manufacturing adjustments.

Versatility and Flexibility: The versatility and flexibility of creating widely different properties of synthetic fibers largely explain why they account for nearly 65% of the worldwide fiber consumption reviewed in **Chapter I**. While there are too many to mention, one relatable example for everyone is the use of synthetic fibers in vehicle safety components such as seat belts and airbags. Although these safety parts could theoretically be made from natural fibers, doing so would be both impractical and dangerous. To achieve the same strength as synthetic fibers, seat belts and airbags made from natural fibers would need to be five to ten times heavier, which would pose significant safety risks from their own weights up on collision. Additionally, natural fibers degrade more rapidly under extreme conditions inside vehicles in hot and cold climates. This would necessitate frequent replacements to maintain their strength and functionality. Clearly, relying on natural fibers for such application does not make sense. There are numerous other examples where natural fibers are neither practical nor beneficial in performance metrics. This consideration naturally leads to the next discussion on the chemical reactivity and toxicity.

Chemical Reactivity and Toxicity

One critical aspect of the environmental impacts associated with different fiber materials is their chemical reactivity and its associated environmental impacts. Although this factor is arguably the most important to consider, it is often overlooked by many environmental interest groups and activists who assert that natural fibers are more sustainable. A common assertion among them is that synthetic fibers are "toxic." However, this characterization does not align with the traditional definition of the term "toxic" or "toxicity".

Definitions of Cambridge Dictionary -

Toxicity: the quality of being poisonous, or the degree to which something is <u>poisonous</u>

Poison: a substance that can make people or animals ill or kill them if they eat or drink it

Overview of Toxicity: In a conventional viewpoint and a layman's term, toxicity can be defined as the property of a chemical to create immediately harmful effects on the health of living organisms. For example, a poison can kill humans instantly at a certain concentration. Common examples of toxic chemicals include heavy metals such as lead, mercury, cadmium, etc., pesticides and herbicides, as well as a wide range of other chemicals found in various industrial and farming applications such as benzene, formaldehyde, etc.

These toxic substances possess chemically active functional groups that readily react with their surroundings. These reactions occur instantly and violently. Although lower densities of some of these chemicals may not pose immediate health impacts, they can accumulate in living organisms over time and result in serious mid- to long-term health effects.

Chemical Reactivity of Natural and Synthetic Fibers: While further details will be explored in the subsections of this chapter, it is important to understand that natural fibers contain numerous functional groups in their chemical compositions that readily react with other elements and substances they come into contact with. In contrast, synthetic fibers either lack chemically active functional groups altogether or have them in much lower densities, resulting in significantly reduced reactivity compared to natural fibers. In fact, this chemical stability of synthetic fibers is a key factor in their lower consumption of chemicals and water throughout their entire lifecycles, from manufacturing to consumption, significantly benefiting the environment.

"Toxicity" vs. (Longer-term) "Damage": Labeling synthetic fibers as "toxic" by some environmental interest groups and activists may stem from studies that show microplastics can accumulate in the body and potentially lead to health issues. However, this is not due to the chemical reactivity, or lack thereof, of the microplastics themselves. Instead, it is the result of other forces that cause certain substances to adhere to or detach from one another. This phenomenon typically involves weaker forces such as surface tension, adsorption and Van der Waals force, which apply to anything entering the bodies of living organisms, even in the absence of chemically active functional groups. Moreover, these forces are much weaker than stronger bonds formed by chemically reactive substances. As a result, they can easily break apart.

Despite the chemical inertness and weak forces involved in the microplastics, potential risks arise when other chemicals adhere to them, eventually forming harmful compounds. In the meantime, the cumulative effects of microplastics and their interactions with other chemicals to create such compounds tend to occur over longer periods, which distinguishes its nature from the traditional concept of "toxicity".

Given these factors, it is more accurate to describe the health impacts of microplastics as (longer-term) "damage" rather than "toxicity". Furthermore, the resulting compounds would not pose significant risks if the adhering chemicals were not toxic. In this context, we understand that the harmful chemicals are the primary sources of "toxicity" concerns, not the microplastics.

Easy Understanding of Chemical Reactivity: The chemical reactivity of a fiber type can be easily grasped without an in-depth knowledge of chemistry. The absorption of water is the result of a chemical reaction between the fiber's chemically functional groups and water. Therefore, one easy way to estimate the chemical reactivity of a material is by observing how much water it absorbs. For example, one can easily test the following at home: submerge two towels, one made of polyester and the other of cotton, in water, then squeeze them to remove excess water. You will notice that the cotton towel retains much more water than the polyester one, which can be further verified by comparing the drying speed of the two in hang dry - I conducted a similar test using my home laundry setup and a weighing scale. The results, which highlight a large difference between cotton and polyester, are detailed in **Chapter III: Section 13. Lifecycle Assessment (LCA) - Consumption**.

In scientific terms, standardized measurements such as "Moisture Content (MC)" are used to show how readily a material interacts with water molecules.

Differences in Moisture Content: The MC is typically measured under controlled conditions, with relative humidity ranging from 35% to 60%, temperatures between 20°C and 25°C (68°F to 77°F), and a stabilization period (e.g., 24 hours) in laboratory settings. The results of such measurements serve as a reference in the trading of fibers and yarns, ensuring that buyers pay for the fiber's weight rather than the water absorbed.

Arrived from these standardized methods, polyester has the moisture content of around 1%, and polypropylene is a mere 0.1%. Although both polymers lack chemically reactive groups in their compositions, they still absorb small amounts of water due to the previously mentioned weak forces such as surface tension, adsorption, Van der Waals, etc., present in nature. In comparison, cotton has the moisture content of 6% ~ 8%, and wool ranges from 16% to 18%.

These differences in the MC reflect significant variations in the chemical reactivity, with natural fibers being tens or even hundreds of times more chemically reactive than synthetic fibers.

Differences in Chemical Absorption: Fibers that absorb more water also tend to absorb higher quantities of process chemicals, leading to increased toxicity throughout their lifecycle. This toxicity is a critical environmental and health concern, which we cannot disregard when we discuss the environmental impacts from different fiber types. In the subsequent sections, I will explore the chemical compositions of different fibers and how they contribute to varying levels of toxicity and overall environmental impacts.

Section 7 Natural Fibers

Natural fibers are derived from two primary sources: plant-based and animal-based. Plant-based fibers include cotton, linen, hemp, jute, etc., while animal-based fibers include wool, silk, angora, alpaca, mohair, etc. Each individual fiber must be long, flexible, and strong enough to withstand manufacturing processes to be suitable for use as textile material.

Different species of plants and animals produce fibers of varying characteristics. For example, garments made from merino wool are known for their smooth texture and natural drape, owing to the fine and soft hair produced by the Merino breed of sheep. Similarly, mohair, which comes from the Angora goat, is renowned for its silky sheen and even softer texture than the merino wool. These differences, though subtle in terms of their chemical composition, are often significant enough for consumers to distinguish between them based on their physical properties.

The characteristics of the fibers are measured using various metrics, including appearance, diameter, length, surface structure, smoothness, strength, moisture content, etc.

Consumption Breakdown: Among natural fibers, cotton has the highest consumption, accounting for approximately 23 ~ 25% of the global fiber use. In contrast, wool, the next most commonly used natural fiber, constitutes only about 1% of the worldwide fiber consumption. Other natural fibers such as silk, hemp, linen, etc., represent negligible proportions of total fiber consumption due to significantly limiting factors such as physical and chemical properties, availability, processing costs, price-quality ratios, etc.

Consumer Preferences: Many consumers would prefer products made from natural fibers over those made from synthetic fibers. This preference can be attributed to various reasons based on some perceived values such as comfort and skin-friendliness of natural fibers, aesthetic appeals and their belief that natural materials are more sustainable. High fashion industry often reflects these consumer preferences, as luxury products are typically expected to be made from natural fibers rather than synthetic ones. Although there are exceptions where synthetic materials are justified, this general preference applies to a significant portion of the textiles people consume, especially everyday items such as shirts, pants, suits, etc.

Of the considerations, the perceived environmental benefits of natural fibers have become a focal point by many environmental advocacy groups, activists and consumers alike.

7.1 Cotton - King of All Natural Fibers

The basic chemical composition of cotton is cellulose, which contains numerous hydroxyl groups (-OH) as shown in *Fig.24*.

Industry Specifics (4) – Start

Fig.24 Chemical Structure of Cellulose*

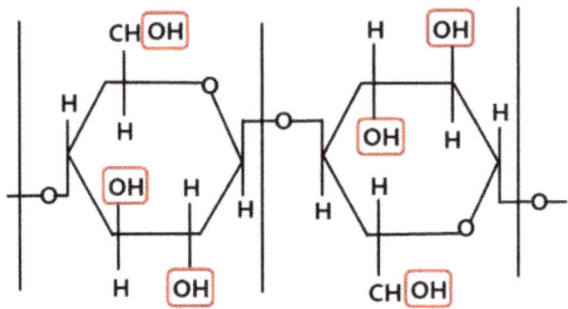

* Hydroxyl groups are highlighted in the red boxes

Although various terms and descriptions exist to express the chemical behaviors and physical properties of the cellulose, the vast majority of them stem from or are governed by the chemical reactivity of the hydroxyl groups. Examples are below:

Hydrogen Bonding: The hydroxyl groups in the cellulose can form hydrogen bonds with other neighboring hydroxyl groups. These bonds impart strength to cellulose, essential for the subsequent manufacturing processes and consumption. Under the same principle, hydrogen bonding also occurs with neighboring water molecules, and it is referred to as hydrophilicity, explained next.

Hydrophilicity: The strong affinity of the cellulose for water is termed "hydrophilicity" (noun) or "hydrophilic" (adjective). This property impacts the entire lifecycle of the cellulosic fibers. For instance, cotton fibers exhibit higher strength when wet compared to when dry. If cotton becomes too dry, it is more prone to breaking and shedding dust. Therefore, this hydrophilicity plays a critical role in cotton (and other plant-based fibers), maintaining its strength during manufacturing and consumption thereafter.

The hydrophilicity of the cellulose also creates other implications. Because the cellulosic fibers readily absorb and retain water, they take longer to dry, leading to increased energy use for manufacturing and laundry during consumption. Additionally, these fibers have a higher probability of shrinking, color-fading and -contamination if not properly handled, especially during laundering. For example, washing a cotton shirt in hot water and using high-temperature dryer settings can result in noticeable color issues and shrinkage. This phenomenon is attributed to the hydrophilicity of cellulose.

Industry Specifics (4) – End

Chemical Reactivity: The hydroxyl groups in the cellulose contribute also significantly to their elevated level of chemical reactivity. As a result, the cellulosic fibers like cotton, linen, hemp, etc., exhibit a strong tendency to interact with other chemicals, including pesticides, fertilizers, dyes, detergents, bleach, etc., during cultivation, manufacturing and consumption. Conversely, a higher tendency of color migration and fading mentioned earlier is attributed also to the heightened chemical reactivity of both the cellulose and the color pigments, readily interacting with water molecules.

Biological Degradation – Critically Erroneous Perception: The hydroxyl groups in the cellulose molecules render cotton susceptible to enzymatic degradation by microorganisms such as bacteria and fungi. While rapid biological degradation is a key argument used by environmental interest groups and activists in favor of natural fibers, it is critical to recognize that, while the cellulose degrades, large quantities of chemical elements and compounds present in textile goods do not. These chemicals, used extensively in farming and manufacturing processes, remain either in the fibers themselves or in nature as they are released into the environment during their entire lifecycles.

For example, the colors in textile goods, regardless of whether natural or synthetic materials, are a sign for the presence of the coloring pigments, along with many other chemicals employed to fixate the pigments within the fiber structures. These chemicals can include strong and highly toxic chemicals such as heavy metals like lead, arsenic, mercury, cadmium, etc., as well as various types of aromatic compounds and other harmful chemicals that are persistent in the environment for extended periods.

Various studies have shown how harmful and long-lasting these substances can be, with some known to persist in nature almost indefinitely, contaminating soil, water, and air, and leading to bioaccumulation in living organisms, including humans even in remote locations from their sources.

***

Therefore, it is a significant and detrimental misconception to assume that natural fibers degrade and return to nature quickly without leaving any environmental footprint. Especially in consideration of how much more chemicals are used for natural fibers and their heightened chemical reactivities, compared with those used in synthetic ones, this oversight critically misleads our society regarding the true environmental impacts between them.

7.2 Wool - King of Animal-based Natural Fibers

As many people know, wool refers to the fibers obtained from various animals, most typically sheep, but also from others like angora from rabbits and cashmere from cashmere goats. There is a wide range of sheep, goats, etc., which produce different qualities of wool, varying in appearance, texture, length, and thickness. Some types of wool are more expensive than others due to their more favorable characteristics for consumers. For example, cashmere is renowned for its extremely soft and smooth touch and is much more expensive than the regular wool from sheep.

Industry Specifics (5) – Start

Wool is composed of a particular protein, known as "keratin", as well as some other chemical compounds such as fat, mineral matters, etc. The keratin plays a critical role in the chemical reactivity of wool fibers, consisting of long polypeptide chains with many different amino acids as shown in **Fig.25**.

Fig.25 Chemical Composition of Wool Fiber*

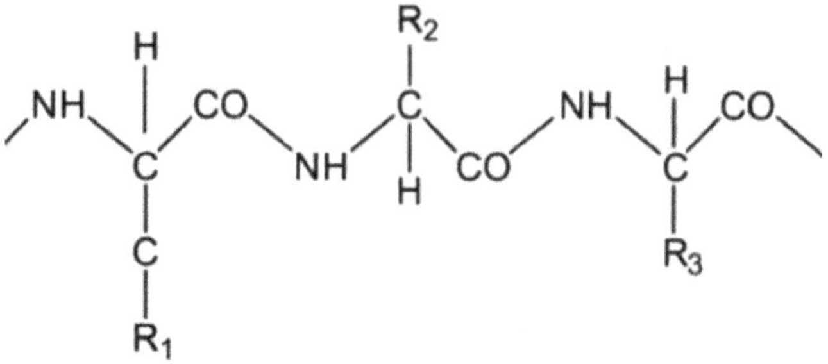

```
          H            R₂                    H
    NH    |    CO      C       NH            |    CO
      \   |   /      /   \       \           |   /
       \  C  /      /     \       \          C
          |       NH       CO                |
          C        |                         R₃
          |        H
          R₁
```

* R side chains of varying characteristics

Some common chemically reactive groups in the proteins include amino group (-NH$_2$), carboxyl group (-COOH), hydroxyl group (-OH), sulfhydryl group (-SH), imidazole group (-C$_3$H$_4$N$_2$), phenol group (-C$_6$H$_5$OH), etc.

Industry Specifics (5) – End

Chemical Reactivity of Wool: As in the hydroxyl group of cotton, the chemically reactive groups of wool, the keratin, make it highly susceptible for chemical reactions with water, harmful chemicals, and other substances. In fact, the proportion of the chemically reactive groups in the molecular compositions of wool is much higher than that of cellulose and it makes wool's chemical reactivity even more elevated.

As outlined in the Introduction of this book, I had extensive hands-on experiences in various textile manufacturing processes including dyeing where I dyed a diverse array of fibers and yarns with varying compositions with my own hands. When attempting to achieve similar tone and hue between wool and polyester, it was necessary to use significantly larger quantities of chemicals. My observations in dyeing simply aligned with the scientific understanding of the chemical compositions of different fiber types and their impacts on chemical usage, thus leading to substantial differences in the overall environmental impacts.

7.3 Other Natural Fibers and Materials

Many other types of natural fibers are utilized in a wide range of textile applications. When combined, the consumption of all natural fibers, excluding cotton and wool, accounts for approximately 5 ~ 5.5% of global fiber consumption. The variations include:

- Plant-based fibers: Hemp, Jute, Linen, Milkweed, Flax, etc.

- Animal-based fibers: Silk, Alpaca, Camel hair, Cashmere, Mohair, Yak, Chiengora, Qiviut, etc.

* **Special Mention:** Additional plant-based fibers, such as bamboo, banana, pineapple, soy, seaweed, etc., also exist. However, these fibers require chemical processing to transform them into certain form factors suitable for textile processing and, therefore, are not classified as natural fibers. This topic will be discussed further in the context of semi-natural, also known as semi-synthetic, fibers later in this chapter.

Leather: Another important form of natural textile materials, albeit not in fiber form, includes leather and fur (subset of leather). Leather, in particular, is widely used in apparel, furniture, automotive parts, etc., and is one of the most commonly used natural materials. Leather is derived from the skin of animals such as cows, horses, goats, sheep, deer, pigs, kangaroos, ostriches, snakes, crocodiles, stingrays, and more. In some cases, leather is used with the hair it comes with, such as in mink, beaver, etc., generally known as fur.

The vast majority of leather, including fur, undergoes extensive processing to enhance its visual appeal, similar to how fibers, yarns, and fabrics are dyed to obtain desirable colors for consumers. This leather treatment, commonly referred to as "tanning," is particularly harmful to the environment due to the use of highly toxic chemicals such as chromium, sulfides, formaldehyde, and reactive dyes. These chemicals are essential for stabilizing the leather and achieving the desired aesthetic and physical qualities, but they come with significant environmental and health risks. Workers in tanneries, as well as nearby residents, may face increased risks of serious health issues, including skin conditions, respiratory problems, and long-term illnesses like cancer, due to prolonged exposure to these toxic substances.

Mycotextile: As briefly mentioned in **Chapter I**, the fungi-based material, known as mycotextile or mycelium, is considered as vegetative leather. There has been a growing popularity on this mycotextile with consumers' general misperception on natural materials as being more sustainable. However, all the previous discussions in the context of the chemical reactivity of natural materials are also applicable for this vegetative leather.

While this material constitutes a minute portion for the overall textile material consumption, it is important to note that farming mycotextile involves the use of sterilization chemicals such as fungicide and carbon footprint from energy use for temperature-, humidity- and light-control, in cultivation as well as methane (CH_4) creation from substrate waste, etc.

Down-Feather

Another important natural material, not in fiber or sheet form, is down and feather material from birds. Certain bird species migrate seasonally, some for breeding and others to escape cold weather. During the migration, these birds require protection against cold air, especially in the chest area because important organs such as hearts, muscles, fat to store energy, etc., are located and must be protected. With this necessity, they developed specialized hair structures in the chest area, known as "down". The down material consists of much finer strains of hair than those of the feathers covering the rest of their bodies. Together, these materials are referred to as "down feather" or simply "down".

Fig.26 Locations of Down and Feather

https://www.sciencelearn.org.nz/resources/2372-whio-feathers-what-are-they-for

The delicate structure of the down creates numerous tiny air pockets to capture and conserve the body heat, a key element for effective insulation. *Fig.27* shows a high quality down material under microscope.

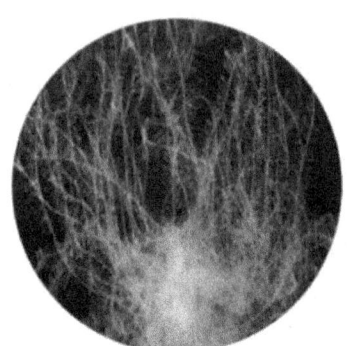

Fig.27 Down under Microscope

When many pieces of this down material are placed in an enclosed bag system for winter jackets, sleeping bags, etc., they form an optimal structure to create a large quantity of extremely small air pockets. The finer the hairs are the smaller and higher quantities of air pockets they can create. Certain species such as the geese from Canada and Hungary are known for producing higher quality down with finer hair structures than others as these birds face colder environment in their migration seasons. Eiderdown is another notable type, recognized for its exceptionally fine hairs, offering superior insulation value.

Value of Down vs. Feather: When the material is used in textiles, it typically consists of a blend of both down and feather. The higher the proportion of the down in the mix, the greater the value of the material. Given the higher importance in its insulative value of the down, it is often referred to as just "down" by many, while "down-feather" would be more descriptively accurate. Although rare, some high-end products contain 100% down, in which case, calling it as "down" would not be erroneous. The quality of a down feather material is often indicated by the ratio of the down to the feather, such as 90/10, meaning 90% down and 10% feather in the blend. Common ratios for apparel, home textiles, sleeping bags, etc., include 90/10, 80/20, and 70/30, where 90/10 material comprising of a high quality down can be multiple times more expensive than 70/30 in a regular quality. In some countries, a material cannot be labeled as down unless the down blend meets a certain threshold, for example, 60%, meaning that the down/feather mix must be minimally 60/40 ratio.

Fill Power: The quality grading of down also involves measuring its Fill Power, which is the volume a specific weight of down occupies. The higher the Fill Power, the more quantity of air it contains, thus the better the quality and the more expensive the material. For example, a Fill Power of 900 indicates a high-quality down for most consumer products, while a Fill Power of 600 is considered lower end.

From my perspective, the Fill Power provides a more objective quality reference than the blend indication between the down and feather materials. For instance, a 90/10 down from a Hungarian goose, typically a strong indication of high quality, yielding a low Fill Power of 600, could suggest potential issues such as wrong material source, improper storage, high moisture content, inaccurate down-feather blend ratios, etc.

***

The down feather material has been widely used for winter goods, with its first recorded use dating back nearly 5,000 years ago. Unfortunately, it is also one of the most misunderstood textile materials. Given its importance on how it impacts our lives from the protective nature of winter goods, it merits a deeper exploration. However, the discussions extend beyond the main themes of this book, thus I reserve this discussion on another book, focusing on the thermal technologies for textiles, expected to be published in the early part of 2025.

Section 8 Synthetic Fibers

Chemical compositions of synthetic fibers vary widely from one type to another. Most commonly used synthetic fibres include Polyester, Polyamide (with commercially known name as Nylon), Polypropylene (also known as Olefin), Acrylic, etc.

8.1 Definition of Plastics

Synthetic fibers are categorized under the family of plastics, specifically thermoplastics. Plastics, in a broad sense, refer to any polymer-based material that can be shaped or molded when heated. According to this definition, natural rubber could be considered a plastic. However, in common usage, the term, plastic, is used for materials synthesized from petroleum sources, thus excludes natural materials like rubber.

There are two primary categories of plastics: thermoplastics and thermoset plastics. Although both types are synthetically produced polymers derived from naphtha, a byproduct of oil refining, they exhibit critical differences in terms of the environmental sustainability.

The thermoplastics can be reheated above a specific temperature known as "Melting Point", unique to each material type. Upon reheating, they revert to a viscous liquid form suitable for reprocessing. In contrast, the thermoset plastics, or thermosets, undergo irreversible chemical changes when exposed to elevated temperatures. Instead of melting, the thermosets typically combust, making it impossible to recycle under the melting principle of the thermoplastics. These fundamental distinctions are pivotal in determining the recyclability of these materials.

Real-Life Applications: Common examples of the thermoplastics include synthetic fibers such as polyester, polypropylene, Nylon, acrylic, etc. Additionally, thermoplastics are widely used in daily items such as plastic bags, water bottles, food containers, etc. In contrast, the thermoset plastics include materials like epoxy resins, which are commonly used in adhesives, paints, and coatings, as well as phenolic resins used in electrical insulators, among other applications. Moreover, composite materials are created by blending thermosets or thermoplastics with other materials to achieve unique properties. For instance, epoxy resin reinforced with fiberglass is used to produce strong, flexible vehicle bumpers that are lighter than traditional steel bumpers. This lightweight, flexible material provides significant benefits, such as reducing impacts on drivers, passengers, and pedestrians upon collision, while also contributing to improved fuel efficiency due to its lighter weight.

Misperception: There are countless innovative applications of plastics and composites that have significantly enhanced our quality of life. For example, hollow synthetic fibers have been used in heart surgery to create artificial veins, which have extended patients' life expectancy and improved their quality of life. Numerous other examples illustrate how synthetic fibers, and plastics in general, have positively impacted our lives. Listing all these benefits in detail would expand this book to thousands of pages, if only feasible.

***

One of the key objectives of this book is to present substantial evidence challenging the widespread belief that plastics are inherently harmful and should be eliminated. The environmental interest groups and activists who advocate for the elimination of plastics may not fully realize the extent to which their daily lives depend on them whether in medicine, food production, health, safety, or even the environmental sustainability. Since this book focuses on textiles, I will explore the crucial role the synthetic fibers play in improving overall human well-being and their significant contributions to the environmental sustainability in the discussions that follow.

8.2 How Plastics are made

As previously mentioned, the plastics are derived from petroleum sources. Some people may refer to the source material as naphtha, while others use the term, crude oil - In certain industries and regions, naphtha is synonymous with crude oil. However, it is more conventional to distinguish between the two, with crude oil serving as the source material and naphtha being one of the "fractions" extracted from it. Naphtha can also be derived from other sources, including natural-gas condensates, fractional distillation of coal tar, peat, etc. This discussion will focus on the crude oil as the most conventional source material.

Industry Specifics (6) – Start

Refinery Process: Crude oil is a complex mixture of thousands of different compounds, which are extracted through a process called "distillation" or "refinery". As shown in *Fig.28*, it involves heating the oil, which then enters the distillation tower.

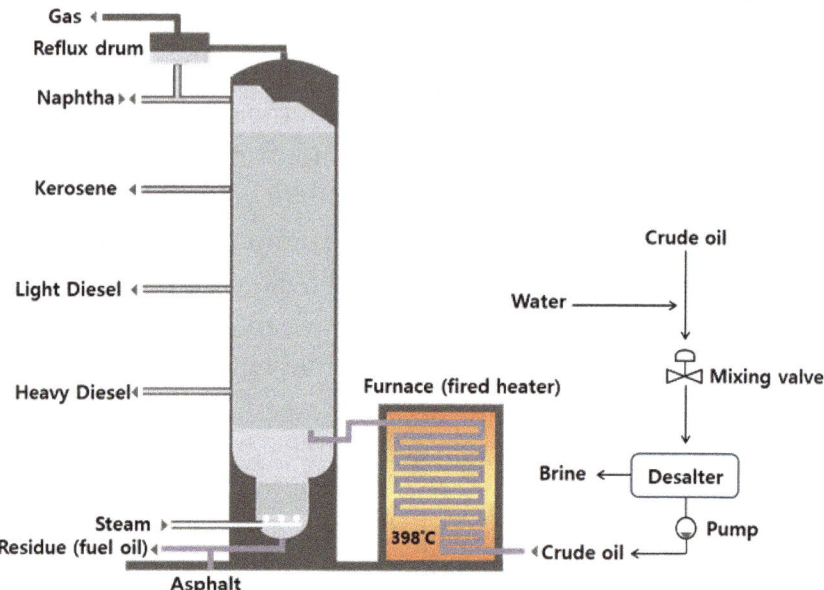

Fig.28 Refinery Process Diagram

Upon heating, the compound with the lightest molecular weight evaporates first and rises to the topmost section of the distillation tower. Successively other compounds follow in order of molecular weight. These separated compounds are referred to as "fractions".

Fractions: Generally known names for these fractions are gasoline, LPG, naphtha, kerosene, diesel distillate, medium- and heavy-weight gas oil, asphalt, etc. This explanation is generalized and does not cover numerous other fractions from refinery processes. For example, there are different grades of gasoline, such as higher and lower octanes, as often seen in gas stations. As this book is not intended to delve deeply into refinery processes and related topics, the basics covered here should suffice to serve its purpose for further discussions on textiles. If there are readers who wish to learn more about it, there are plenty of information available on various sources.

Pricing Factors: The market price of crude oil depends on the proportion of higher-value fractions, such as gasoline, compared to lower-value fractions, such as asphalt, and whether the oil has higher or lower content of sulfur, referred to as either "sour" or "sweet" oil. For example, the West Texas Intermediate (WTI) and the Brent oil are classified as light oils because they contain higher proportions of valuable fractions in lighter weights. In addition, they contain lower amount of sulfur in a given weight, making them sweet oil. These two characteristics make them more valuable in the market. In contrast, the Maya crude from Mexico contains more heavy compounds like asphalt and higher content of sulfur. It is with this reason that the Mayan crude oil is called heavy and/or sour oil and is generally lower in price.

To capture the fractions effectively, the distillation towers are typically quite tall and characteristic of refinery facilities. Many readers may have seen such towers while driving through the industrial areas of large cities.

Industry Specifics (6) – End*

8.3 How Synthetic Fibers are Made

All synthetic fibers are subsets of plastics and are often referred to as "polymer" in a more scientific context. The word "polymer" originates from Greek, where "poly" means many and "mer," derived from "meros," means parts or

167

units. Thus, "polymer" signifies a structure composed of many repeating single units. Other names for these materials include thermoplastics and man-made fibers. "Monomer" ("meros"), is the term indicating the basic repeating unit when it used independently. A monomer is a molecule capable of chemically bonding with other identical molecules to form a polymer. Monomers are typically in a liquid state at room temperature whereas polymers exist as solids.

Various types of polymers are made from different monomers, such as esters for polyester, amides for polyamides, propylene for polypropylene, and acrylates for acrylic.

Industry Specifics (7) – Start

Polymerization Principles: To create chemical bonding among monomers in a liquid form, specific processing conditions are required. These conditions may include such factors as temperature, pressure, catalyst, etc. When the conditions favor the synthesis of monomers, polymer structures form. Typically, these formations occur at high temperatures and in linear directions due to the chemical bonding structures of monomers, a concept elaborated in *Fig.29*. This process is generally referred to as "polymerization".

Fig.29 Polymerization

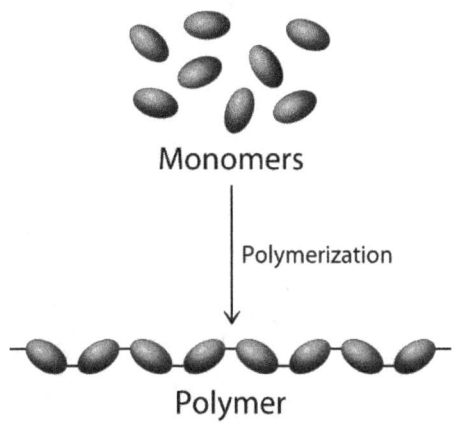

This polymerization occurs in a metallic tank where the conditions mentioned earlier can be optimally adjusted. This equipment is generally referred to as "polymerization tank".

Extrusion Principles: In the polymerization tank, the polymer chains are randomly positioned and intertwined in a viscous liquid form. To convert this into long strands of fibers, the liquid is pushed out of the tank through small holes as illustrated in **Fig.30** and is subsequently pulled by a winding mechanism at the end of the process. As the processing conditions outside the tank are cooler than inside, the liquid cools and solidifies. This process is known as "extrusion".

Fig.30 Polymerization and Dry Extrusion

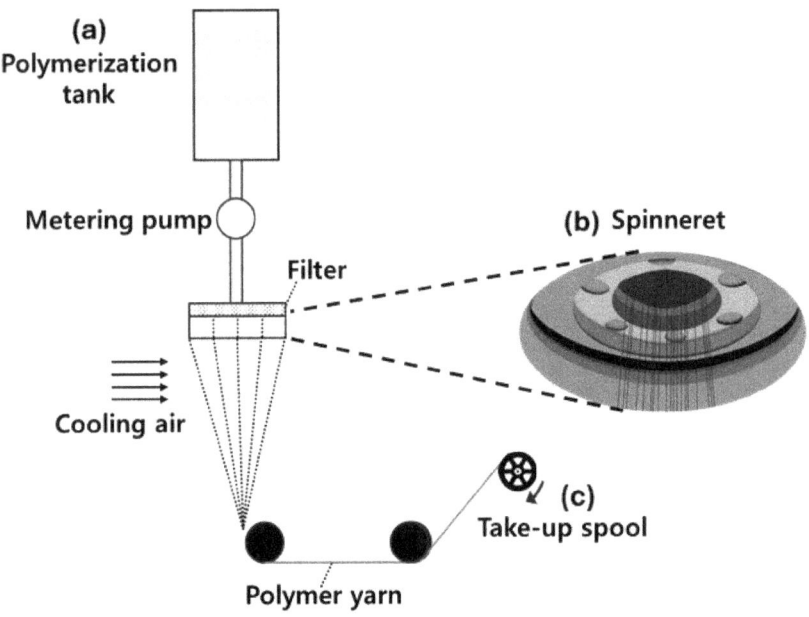

For example, in the case of PET, a subset of polyester, the temperature at which the polymer exists in a viscous liquid, known as the Melting Point, is between 250°C(482°F) and 260°C(500°F). There is a temperature range between 220°C(428°F) and 240°C(464°F), known as the Softening Point where the polymer is in a state between liquid and solid. Then, the PET solidifies into fibers below 220°C(428°F). As the air outside the melting tank is much lower than the solidification temperature, the strands quickly cool and become the fibers.

When a polymer in a molten state is extruded from the polymerization tank into ambient air, the process is referred to as "dry" extrusion and **Fig.30** is an example. When a polymer is extruded into a chemical bath, it is known as "wet" extrusion. The choice between these two methods typically depends on polymer type, solidification efficiency, desired quality of resulting fibers, etc. The polyester fibers are generally extruded through the dry extrusion, while the acrylic fibers require a bath of acidic chemicals, thus employing the wet method.

Other Plastic Formation Principles: Principles of creating other form factors, such as the sheets for plastic films or the bottles in tubular one-end-enclosed shapes differ mainly in this extrusion process. For instance, to produce a plastic film from the melting tank of the ***Fig.30*** process, the viscous polymer must be laid onto a flat surface and solidified into a desired thickness. Instead of the small holes used in the fiber extrusion, the mechanical setup allows the formation and solidification of the sheet form factor at the end of the polymerization tank. Similarly, the plastic bottles are the result of the viscous polymer poured into a mold shaped with a desired form factor, size and thickness during the extrusion process.

Industry Specifics (7) – End

Variations and Versatility: Differences in synthetic fiber types are determined by the chemical compositions and structures of each polymer type. As in the natural fibers, the synthetic fibers are characterized by similar metrics such as appearance, thickness, length, surface structure, texture, strength, etc. As mentioned earlier, the physical characteristics of the synthetic fibers can be easily adjusted by tweaking manufacturing parameters. For example, the thickness of the synthetic fibers can be adjusted by changing the hole sizes for the extrusion process or modifying the speed of the extrusion. This is akin to making thinner or thicker pasta noodles from flour dough by adjusting the thickness settings of pasta machine. The ability to make such fine adjustments allows for the versatile creation of different fiber characteristics, significantly impacting the properties of final products.

With this general overview, let's now delve into the specifics of some of the most conventional types of synthetic fibers.

8.4 Polyester, King of All Fibers

Polyester is arguably the most important fiber, accounting for over 50% of total fiber consumption. Given hundreds of different fiber types available to us, including natural fibers such as cotton, wool, silk, mohair, alpaca, cashmere, etc., as well as other synthetic fibers like Nylon, acrylic, polypropylene, aramid, carbon, etc., the fact that the polyester single handedly constitutes half of the global fiber consumption is significant and indicates clear advantages this fiber offers, including strength, softness, durability, flexibility, low manufacturing cost, ease of maintenance, chemical resistance, among many others.

Usage Overview: Thanks to these excellent attributes, polyester's applications are remarkably diverse, encompassing both personal items such as apparel, home furnishings, sports gear, etc., and industrial applications like tire reinforcement, safety belts, hard hats, spray coatings, etc. As mentioned earlier, hollow polyester fibers are used in the heart bypass surgery to carry blood around the body, and polyester fabrics are employed in surgeries requiring implantable supports in our body. The polymer is also used in stents, bone fixation devices, and numerous other important medical applications.

Chemical Inertness: Polyester's suitability for the medical applications stems from its chemical inertness, meaning it does not react with other chemical elements and compounds. In comparison, if natural fibers and fabrics were used in such applications, they would dissolve in the body, as they are chemically reactive, leading to critical health issues in a short period post-surgery.

Industry Specifics (8) – Start

Chemical Composition: The monomer of polyester is comprised of the ester group, hence the name, Polyester, i.e., many ester groups. A specific type of polyester is polyethylene terephthalate (PET), and it is a material quite often spotted on in the care labels of various textile goods. The popularity and importance of this fiber type come from the favourable molecular structure in its chemical composition.

Fig.31 Chemical Structure of Polyethylene Terephthalate (PET)

$$H-(OCH_2CH_2O-\overset{O}{\underset{\|}{C}}-\underset{}{\bigcirc}-\overset{O}{\underset{\|}{C}}-)_n OH$$

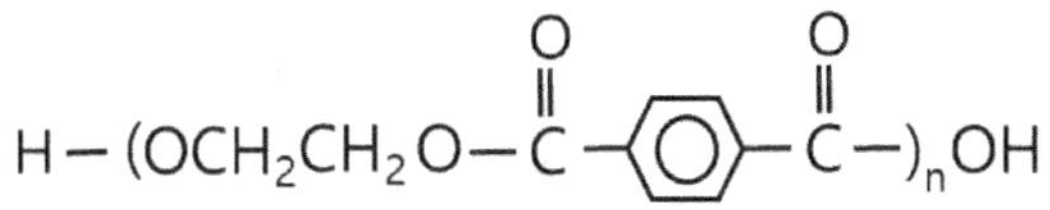

Industry Specifics (8) – End

Characteristics: The fiber's molecular structure is stable, and it does not react to other elements and chemical compounds. Low absorbance of water is a phenomenon representing such characteristic. Since water is a key source of a wide range of issues in manufacturing, consumption and even disposal, including bacterial growth causing odors and health issues, shrinkage, color migrations, contamination in landfills, etc., the benefits sub-branched from the low level of affinity to water are numerous.

Environmental Benefits and Disadvantages: Among them are environmental benefits. For example, it uses much less water in manufacturing and laundry during consumption in comparison with natural fibers. In addition, it absorbs much lower quantity of other chemicals, including toxic ones, during its lifecycle. Textile goods made with 100% Polyester do not require drycleaning, in most cases, consume less water and energy in home laundry and need little to no ironing.

There are numerous other important benefits the polyester present for the environment. However, the polymer also presents significant challenges such as longevity in nature and microplastics, highlighting the importance of comprehensive analysis for assessing the true environmental impacts in comparison with other fibers. These factors will be explored in depth throughout this book including **Chapter III: Lifecycle Assessment (LCA)**, where the 16-point evaluation criteria I developed with a focus on analysing the factors as comprehensively and objectively as possible will be utilized.

Versatility: The polyester is also one of the most widely used fiber types blended with other fibers. For instance, cotton and polyester blend is one of the most common types of yarns and fabrics while numerous other blends with natural fibers or other synthetic fibers popularly exist. These blends are to create a certain set of characteristics which take advantage of the benefits offered by each blended fiber type. For example, the cotton and polyester blend reduces the risks of shrinkage, color fading, ironing need, etc., thanks to the chemical inertness of the polyester, while providing natural texture and higher moisture contents close to the skin from the cotton. The trade-offs resulted from different blend proportions between two or more fibers are some of the key considerations in engineering yarns and fabrics in the textile goods we consume.

***

The polyester's exceptional properties, along with its relatively low manufacturing costs and widespread use across nearly every aspect of our lives, have earned it the title of the "King of all fibers." Its significant environmental benefits will be examined further throughout the remainder of the book.

8.5 Nylon, Mother of All Synthetic Fibers

Nylon is the commercial name of the polyamide with the amide group as the repeating monomer unit. It is the very first invention in the family of the synthetic fibers, by brilliant scientists at DuPont at the dawn of the 20th century. Today, the fiber accounts for approximately 5% of the global fiber consumption. Nylon is renowned for its strength, smooth texture, and silk-like sheen. When the stocking made of the fiber was commercially introduced in 1940, which had been previously made with silk or cotton, the product was an instant commercial success, selling over 60 million pairs in their first year on the market, despite being more expensive than silk or cotton stockings at the time.

As in other synthetic fibers, it is a thermoplastic, meaning that it can be converted back into the molten state and repolymerized for recycling. There are different chemical compositions in the family of Nylon and Nylon 6 and Nylon 66 are some of the variants.

Industry Specifics (9) – Start

Fig.32 Chemical Structure of Nylon

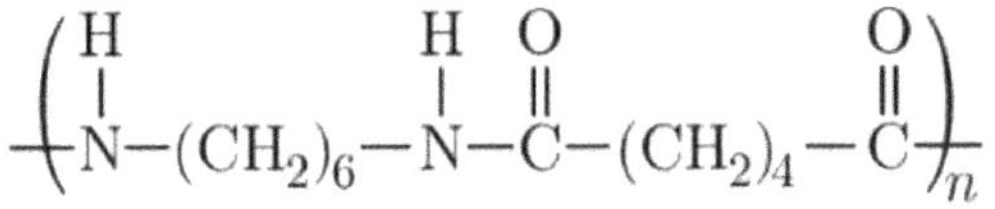

$$\left(\underset{\underset{\mathrm{H}}{|}}{\mathrm{N}} - (CH_2)_6 - \underset{\underset{\mathrm{H}}{|}}{\mathrm{N}} - \underset{\underset{\mathrm{O}}{\|}}{\mathrm{C}} - (CH_2)_4 - \underset{\underset{\mathrm{O}}{\|}}{\mathrm{C}} \right)_n$$

Nylon 66

$$\left(\underset{\underset{\mathrm{H}}{|}}{\mathrm{N}} - (CH_2)_5 - \underset{\underset{\mathrm{O}}{\|}}{\mathrm{C}} \right)_n$$

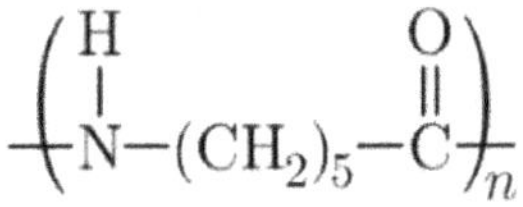

Nylon 6

Industry Specifics (9) – End

Like polyester, Nylon is used in a wide range of applications, both personal and industrial, including apparel, toothbrushes, automotive parts, food packaging, fishing nets, resin for 3D printing, tubes, films, etc.

While both Nylon and polyester belong to the family of synthetic fibers, Nylon is more chemically reactive than polyester under certain conditions. This difference arises from the chemical composition of their monomer units with the amide functional group with nitrogen – carbon bond, which can undergo chemical reactions such as hydrolysis, etc.

As a result, Nylon typically has a moisture content in the range of 3 ~ 4%, while polyester has around 1%. In comparison, cotton has a moisture content of 7 ~ 8%, and wool has 16 ~ 18%. As explained earlier, the moisture content can be roughly translated to higher consumption of water and chemicals during manufacturing and consumption.

While there may be more areas of interests for readers in Nylon, its relatively small consumption rate prevents me from delving deeper into other details. As a general rule of thumb, however, the environmental impacts of Nylon can be considered with its position between polyester and cotton assimilated by the different moisture contents described earlier.

8.6 Polypropylene, Rising Star

Polypropylene is a polymer made from the monomers of propylene. It belongs to the chemical group known as polyolefins and is often referred to as simply "olefin". The invention and commercialization of the polymer occurred in the 1950s and it exhibits excellent impact and fatigue resistance properties among many other advantages. The polymer is commonly found in such everyday items as food containers, pharmacy prescription bottles, chairs, clear bags, stationery folders, etc., in addition to apparel items. The fiber forms of polypropylene are used in face masks, thermal insulation, diapers and sanitizers, as well as in many industrial applications.

Industry Specifics (10) – Start

Fig.33 Chemical Structure of Polypropylene

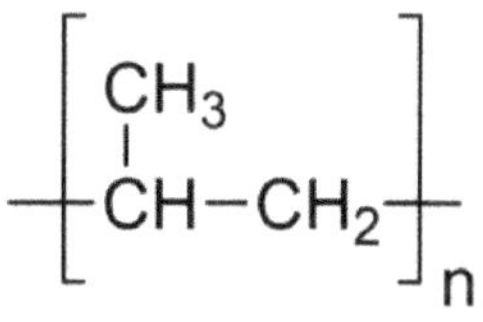

$$\left[\begin{array}{c} CH_3 \\ | \\ CH-CH_2 \end{array} \right]_n$$

Industry Specifics (10) – End

I used the expression of the "***Rising Star***" as the title of the subsection, because the textile business world starts realising the versatility of it not only in the excellent physical properties but also in the environmental benefits.

Of all synthetic fibers of large commercial uses, polypropylene offers the lowest density of polymer structure. The low density offers several significant advantages. For instance, it has lower softening and melting points compared to polyester, leading to requiring lower heat and energy consumption during its manufacturing and consumption. This characteristic also contributes to the creation of lighter products. However, the lower density and melting points can lead to easier deformation, particularly when exposed to high heat in direct contact. Consequently, the industry typically avoids using the fiber for items that are in direct contact with high heat above 100°C (212°F) such as those commonly ironed like dresses, shirts, pants, etc.

Polypropylene is one of the most environmentally beneficial materials of all fibers due to its extremely high chemical inertness. For instance, its moisture content is approximately 0.1%, only one tenth of that of polyester. In comparison, cotton has a moisture content of 7 ~ 8%, and wool has 16 ~ 18%. Therefore, the fiber incurs tens or hundreds of times lower chemical reactivity, benefiting the environment in related aspects. Personally, I have a strong affinity for this material, as it inspired me to develop the technology platform of the **Clean Recycling Initiative**™ - more details provided in **Chapter V: Solutions**.

8.7 Other Synthetic Fibers

There are numerous other types of synthetic fibers including acrylic, spandex, kevlar, carbon fiber, etc., which together comprise approximately 5% of global total fiber consumption. These fibers exhibit distinct differences in their chemical compositions as well as physical and chemical characteristics. Given their relatively insignificant proportions in the total fiber consumption, a detailed examination of each is beyond the scope of this book. Rather, I present several names of different fiber types with typical application examples:

Acrylic: sweaters, transparent boards, etc.

Modacrylic: artificial wigs, flame retardant garments, etc.

Carbon fiber: golf clubs, tennis rackets, helmets, etc.

Polyaramid: bulletproof vests, flame resistant garments, etc.

A notable commonality among these fibers is their origin in the petroleum bases, which imparts molecular structures with significantly lower chemical reactivity compared to natural fibers. While there are variations in the chemical reactivity among different synthetic fiber types, differences within the synthetic group are generally much more subtle than those observed between synthetic and natural fibers.

Section 9 Semi-Synthetic Fibers, also known as Semi-Natural Fibers

Semi-synthetic fibers are a group of fibers derived from natural materials such as wood pulp, bamboo, soy, corn starch, etc. The chemical structures of these natural ingredients are converted through various artificial re-polymerization processes using different chemicals. Typically, this involves various physical or chemical means of, for example, dissolving the main ingredients in a strong solvent chemical to create a viscous solution or turning them into powders for forming desired shapes using various types of adhesives to bond them together. Examples of the later include beverage straws and cups made of natural ingredients instead of other materials like plastics, metals, ceramics, etc. In the applications of textile materials, it is invariably the dissolution method used.

The polymerization process for the semi-synthetic fibers is similar with that of the synthetic fibers particularly in the mechanics of the polymerization and extrusion: Long chains of polymers are formed in viscose liquid states under specific manufacturing conditions, then extruded through spinnerets with small holes to create linear fiber shapes. The main difference between the two fiber types is that the synthetic fibers require primarily heat energy for the polymerization of monomers with little to no intervention from other chemicals while the semi-synthetic fibers involve chemically strong solvents to dissolve the naturally sourced materials.

Examples of the semi-synthetic fibers include rayon, modal, lyocell (branded as Tencel), bamboo, seaweed, etc. These fibers are typically known for their soft, silky and smooth textures and appearances. A more recent addition to the market is the Sorona, which utilizes corn starch as the source material. Many people perceive the fibers in this category as more sustainable than the synthetic fibers, and they are favored by many environmental interest groups and activists.

However, there are critical perspectives that may be missing in these considerations. For instance, the use of the toxic solvents to dissolve the cellulose-based ingredients leaves serious toxicity footprint for the environment, not to mention that it requires large amounts of water and energy use to wash the toxic chemicals off the fibers before the subsequent manufacturing processes and consumption. Moreover, they possess the characteristics of consuming and releasing large quantities of chemicals like natural fibers and generate microplastics like synthetic fibers.

9.1 Regenerated Cellulose Fibers

The discovery and initial development of the regenerated cellulose occurred in the late 19th century when scientists found that the cellulose from various plant materials could dissolve in specific chemicals. Upon re-solidification, the material exhibited a silk-like appearance, soft and smooth texture, which led to it being initially named "artificial silk" or "wood silk". Driven by the favourable characteristics, it has been used for a wide range of consumer textiles such as shirts, dresses, jackets, etc.

Industry Specifics (11) – Start

Process Descriptions: The general description of the manufacturing process involves harvesting the cellulose, the primary component of plants, which is found in the trunks, barks, roots, and other parts of trees. This cellulose is then subjected to a chemical bath for dissolution. Breaking down the cellulose into a pulpy and viscous solution requires the use of highly intensive chemicals defined as "solvents". The types of the solvents used in this process include the commonly utilized solution of the caustic soda and the carbon disulphide, as well as the amine oxide, the later of which is specifically used in the manufacturing of the Tencel.

Depending on the types of wood sources, chemicals, and processes used for dissolving and processing the cellulose, these fibers are commercially known as Rayon (or Viscose), Modal, and Tencel (or Lyocell). The cellulose contents in the regenerated fibers can vary, typically ranging from high 80% to high 90%, depending on the specific processes employed. The remaining percentage is composed of various residual chemicals, impurities, and additives that result from the source materials and the production processes. The level of the residual chemicals and impurities are considerably higher in the regenerative cellulose fibers than that of the synthetic fibers, which normally ranges less than 1%. This is attributed by the naturally sourced materials containing larger amounts of impurities as well as inherently higher chemical reactivities in them compared with the petroleum sources for the synthetic fibers.

Industry Specifics (11) – End

Development History: Rayon was the first developed in the category of the regenerated cellulose fibers. Subsequently, two other variations of chemical synthesis, using different source materials and solvents, led to the creation of the other fiber types, commercially known as Modal and Tencel (or Lyocell).

The development of Modal was primarily driven by the need to improve the wet strength of Rayon during the 1950s and 60s. Garments made of Rayon were prone to stretching or creasing when laundered with water or exposed to rain, due to the fiber's low wet strength.

A Japanese company, the inventor of Modal, significantly improved the fiber's wet strength by using specifically the beech wood instead of other plant species. This innovation retained other desirable properties of Rayon, such as soft and smooth texture and silk-like appearance, leading to Modal's popularity shortly after its commercialization. I recall purchasing a couple of shirts made of Modal in the early 2000's and appreciating their characteristics. Additionally, Modal was recognized as a more sustainable option as the beech wood requires minimal water for growth and is resistant to pests and diseases, reducing the need for irrigation, pesticides, and fertilizers. With lower water and chemical footprints, along with the improved wet strength, Modal seemed like an ideal solution.

Despite the Modal's advantages, however, its overall environmental impact remained significant, as a large portion of the solvent used in the manufacturing process could not be reused. To address these environmental concerns, an Austrian company, Lenzing AG, developed a process that reclaims and reuses nearly all the solvent used in the process, significantly reducing the chemical footprint on the environment. This new generation of the regenerated cellulose fiber is called "Lyocell", branded as "Tencel", and its production began in the 1990s.

According to Lenzing AG, 99% of the solvent is reclaimed and reused. Given the high volatility and toxicity of the solvents, reclaiming 99% is a notable achievement. Moreover, the solvent used for Tencel, amine oxide, is recognized as less toxic than those used for Rayon and Modal, as it does not produce neurotoxic carbon disulfide.

Although the amine oxide is more expensive than the solvents used for Rayon and Modal, the overall production cost of Tencel is likely lower due to the high reclamation rate. This makes Tencel a prime example of the environmental sustainability not necessarily being more costly. I have long advocated for eco-friendly materials, products, and technologies that do not entail higher costs. It has been shown that only a small portion of the population is willing to pay a significant premium for eco-friendly products, while the majority will not, regardless of their expressed values. More in-depth discussion on this subject is provided in **Chapter V: Solutions**.

Environmental Impacts: The solvents used in the production of the regenerated cellulose fibers are extremely harmful and can cause significant disruption to the ecosystems if released into the environment. In the production of Rayon and Modal, it has been reported that up to 50% of the solvents cannot be recycled and are often released into the environment. This reality makes the chemical processing of these popular soft fabrics particularly detrimental to the planet.

Moreover, the regenerated cellulose fibers retain inherent chemical reactivity derived from the cellulose in all plant-based fibers. Although the chemical compositions of the fibers are slightly altered during their chemical synthesis, as illustrated in **Fig.34**, the presence of the hydroxyl groups remains abundant in the regenerated fibers. Additionally, the introduction of the xanthate functional group, highlighted in red in **Fig.34**, further increases their chemical activity. As repeatedly mentioned, higher levels of chemical reactivity correspond to higher levels of toxicity in the entire lifecycles of the materials in question from manufacturing to disposal, lingering in nature even long after disposal.

Industry Specifics (12) – Start

Fig.34 Xanthation of Cellulose

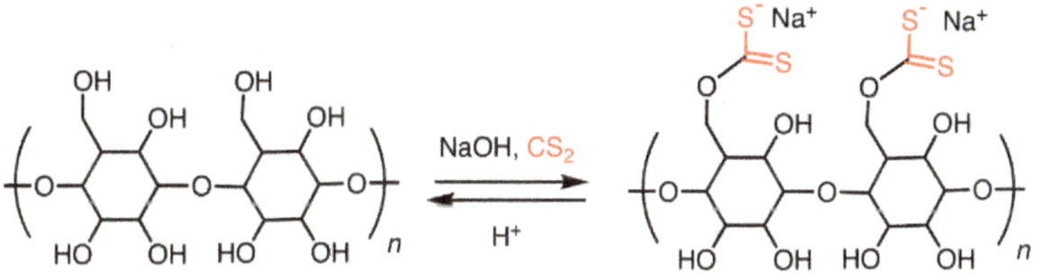

Industry Specifics (12) – End

When environmental groups and activists claim that semi-synthetic fibers are more environmentally friendly than fully synthetic fibers due to their naturally sourced materials, they often focus solely on the carbon aspect associated with the synthetic fibers from their petroleum sources, overlooking the significant environmental impacts of the toxic solvents and the heightened chemical use in this category of fibers. If we were to replace all synthetic fibers with the regenerated cellulose fibers, the resulting toxicity from the solvents alone could cause substantial and immediate environmental damages.

9.2 Sorona®

Sorona® is a different breed of semi-natural fibers, partially derived from corn starch and partially from petroleum-based materials, avoiding the use of solvent unlike other plant-based semi-natural fibers. It was developed by the renowned DuPont company, which has made substantial contributions to the textile industry over the years. In my view, DuPont has enriched our lives more than any other company in the textile sector. The company pioneered the development of several important synthetic materials and technologies widely used in the industry, beginning with the invention of Nylon, the first synthetic fiber, in 1935. This innovation has inspired numerous scientists and companies to develop many other critically important synthetic materials.

The company was awarded the 2003 Presidential Green Chemistry Challenge Award for their development of Sorona. In my opinion, if I had the authority, I would have certainly bestowed upon DuPont the highest possible award for sustainability, in recognition of the groundbreaking Nylon polymer. Thus, it seems to me that this award was granted almost a century late.

Industry Specifics (13) – Start

Process Descriptions and Compositions: Sorona®, known scientifically as poly-trimethylene terephthalate-co-1,4-cyclohexanedimethylene terephthalate, is a polymer that blends plant-derived polymers, specifically corn starch, with petrochemicals. It is known for its flexibility, stretch recovery, smooth and soft hand feel, among other attractive features.

Corn crops are utilized to extract sugar or glucose. Microorganisms are then introduced to ferment the glucose, replacing the need for chemical synthesis and resulting in the natural production of PDO (1,3-Propanediol). This Bio-PDO is then combined with TPA (terephthalic acid) sourced from petroleum processing, forming a molecular linkage and resulting in the creation of the Sorona® polymer.

The natural component from the corn, i.e., PDO, accounts for 37% of the mass of Sorona® while the rest is of the TPA from the petroleum. Therefore, two thirds of the composition is from the petroleum source while many may mistakenly think it is purely from the natural source of corn.

Fig.35 Molecular Structure of Sorona®

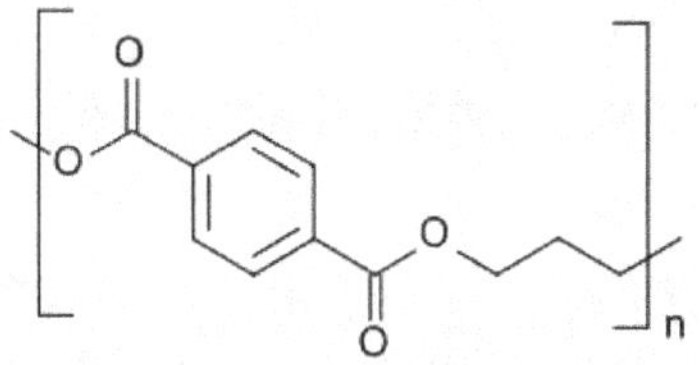

Industry Specifics (13) – End

Social and Environmental Considerations: From a different perspective, using corn as a source material for Sorona® raises concerns, as the corn is a crucial food source for humans and animals alike. Given that one in nine people globally do not have enough food to consume and are malnourished, the sustainability of using a food source to produce fibers is questionable from a humanitarian standpoint. While there are wealthy individuals in developed economies who do not worry about their meals, a significant proportion of the world's population is not as fortunate. These underprivileged would most certainly disagree with the practice of converting food sources into fibers, viewing it, in my view, as unworthy of the 2003 award recognition for the company.

Furthermore, the decomposition behavior and microplastic generation of Sorona® are not markedly different from other synthetic fibers. Sorona's chemical composition, characterized by multiple carbonyl groups in carbon-oxygen double bonds, is stable and exhibits low chemical reactivity. This stability means that Sorona® has a similar lifespan to other synthetic fibers in nature.

The question remains: what justifies the 2003 award as a significant achievement? Is replacing one third of petroleum with corn as the source material sufficient to deem Sorona more "sustainable"? Keeping in mind that the avoidance of toxic solvents is clearly more sustainable than the manufacturing methods employed for the regenerative cellulose fibers, we must also recognize that all synthetic fibers share the same manufacturing principles, not needing solvents.

The overarching theme of this book is that synthetic materials are not inherently detrimental for the environment, and their benefits outweigh the negatives. Had I been in a position to decide, I might not have awarded Sorona for its invention of Sorona®. In my view, DuPont deserved the highest accolades nearly 100 years ago for inventing Nylon, but Sorona® does not warrant the same level of recognition, particularly in the perspective of the "Green" Chemistry Challenge Award, while I have no opposition for such recognition without the "Green" decoration.

9.3 Other Semi-Synthetic Fibers – Bamboo, Pineapple, etc.

A diverse array of other semi-synthetic fibers exists, and growing public environmental awareness has prompted increased interest in these materials although the usage has remained fractional in the global total fiber consumption. Examples include bamboo, soy, pineapple, seaweed, etc.

Similar to the previously mentioned regenerated cellulose fibers, the environmental impacts of these fiber types can be significantly negative with the use of strong solvents. To maintain the focus of this book on the most impactful and important fibers for assessing their environmental impacts, I do not delve into more details of these fibers. Instead, I present their names along with a couple of application examples:

Acetate: cigarette filters, furniture fillings, etc.

Bamboo: pillow covers, towels, etc.

Soybean: socks, curtains, etc.

Seaweed: skin moisturizing and nourishing sheets, towels, etc.

Section 10 Manufacturing

This chapter will examine the manufacturing principles of textile production and their implications for the environmental sustainability. The subsections will be organized in the following subjects: farming for natural fibers, synthesizing for synthetic fibers, yarn-making, fabric-making, dyeing, cut-and-sew, and other, representing a typical order from the beginning to the final stage of manufacturing before reaching consumers.

In the meantime, a detailed understanding of the intricate processes involved in textile manufacturing may not be essential for addressing the important purposes of this book or for comprehending the broader environmental impacts of different textiles. With this in mind, this section focuses on providing a concise overview of the typical flow of textile production. While this approach may make it challenging for some readers to fully grasp certain specific subjects, those seeking a deeper understanding can benefit from further study. For most readers, however, the terminologies and basic concepts introduced here will be sufficient for following the discussions throughout the book and for engaging in related conversations.

10.1 Fibers

In this discussion, we explore the origins of the three main categories of fibers particularly in the manufacturing perspective, which include farming for natural fibers.

10.1.1 Farming

A brief review of the related discussions previously presented: Natural fibers are derived from farming plants that naturally produce fibers, such as cotton, or farming animals that grow hair, such as wool, while leather and down feather is the byproduct of animal farming. Cotton is the most widely used natural fiber and is second only to polyester in overall usage.

Cultivation

Cotton cultivation requires specific agricultural conditions, typically frost-free environments with daily temperatures ideally ranging from 20 to 30°C (68 to 86 °F) and ample sunlight. The world's major cotton-producing countries include United States, Uzbekistan, China, India, Brazil, Pakistan, Turkey, etc. Cotton occupies a significant portion of some of the most prolific agricultural lands around the world.

Thirsty Crop: Cotton is known as a *"thirsty"* crop and the water footprint of it is among the largest of all crops either for personal and animal food consumption or any other purposes. On average, cotton cultivation requires 8,000 to 10,000 liters of water per kilogram of fiber produced. This compares with 3,000 to 5,000 liters for rice and 500 to 4,000 litres for wheat.

These water consumption figures vary widely by various factors including weather conditions and agricultural practices. As such, the quoted figure of 8,000 ~ 10,000 liters for cotton can further increase in hot climates. This excessive water consumption presents significant challenges in both environmental and humanitarian perspectives - Water scarcity is a critical issue affecting the health of humans, animals and other plants in many areas of cotton cultivation.

Water Contamination: Beyond the sheer volume of water used, cotton irrigation practices also lead to significant contamination in water with toxic chemicals such as pesticides and fertilizers. As mentioned earlier, one of the most serious humanitarian crises our global society face today is a significant lack of fresh water in many underdeveloped and developing countries. Highly toxic chemicals used in farming reach wide areas of neighboring regions through runoff water and contamination in soil, leading to pollution in ground water, and further exacerbating the water scarcity.

Toxic Elements and Compounds: Cotton cultivation poses particularly increased environmental challenges due to the toxicity of the chemicals used, many of which recognized as hazardous by the World Health Organization (WHO). A couple of examples of the pesticides include:

Aldicarb: One of the most toxic pesticides applied to cotton. A **powerful nerve agent**.

Endosulfan: One of the most widely used pesticides for cotton. Linked to comas, seizures, and even death. Believed to be the **most important source of fatal poisoning among West African cotton farmers**.

Basic Chemistry: The toxicity of harmful chemicals such as pesticides are partly driven by the containment of heavy elements, which we learned in the periodic table of high school chemistry, e.g., #80 mercury, #82 lead, etc., and other harmful compounds. These harmful chemicals remain in nature, accumulate and magnify within food chains, a process known as bioaccumulation and biomagnification, subsequently being ingested by living organisms. Some of these chemicals can persist in nature for extended period, easily exceeding lifetimes of humans and animals. Notably, Dr. Clair Patterson, whom I will highlight as "GREEN HERO" in **Chapter V: Solutions**, used a specific lead isotope to calculate the Earth's age as approximately 4.5 billion years, as the element endures in nature almost indefinitely.

Toxicity Migration and Bioaccumulation

A significant amount of harmful chemicals, once washed away from one material, do not disappear, but instead, migrate to other materials, either in their original forms or the byproducts of their chemical reactions, through intermediary carriers such as water, air and food sources. This phenomenon results in widespread and persistent impacts on the health and safety of all living organisms on Earth.

To illustrate this issue, I present a report issued by National State of Environment Report, Ministry of Ecology, Environmental Protection and Climate Change of Republic of Uzbekistan and International Institute for Sustainable Development (IISD), in December 2023.

https://www.iisd.org/system/files/2024-02/uzbekistan-state-of-the-environment-en.pdf

Cotton's Contribution for Environmental Harm: The report offers a comprehensive analysis of key factors contributing to the degradation and desertification of the agricultural lands in the country, alongside the drying of the Aral Sea. Across its extensive 120-page length, it depicts the cotton cultivation as a primary factor influencing the current environmental conditions, ranking ahead of the textile industry and other agricultural crops.

Uzbekistan, along with its neighboring Kazakhstan, has played a significant role as a major exporter of high-quality cotton in competitive prices. As the cotton industry has been a cornerstone of Uzbekistan's economy for centuries, with some characterizing the country's economies as "one-crop" economy, this historical emphasis on cotton helps contextualize its prominent listing in the report as the leading cause of numerous environmental challenges facing the country, including agricultural land degradation, water contamination, and even impacts related to climate change.

Nature of Pollution through Air: Upon reviewing the 2023 report, I find it both interesting and alarming that the air quality, expressed as the Maximum Permissible Concentration (MPC) levels in the Atmospheric Air Quality analysis for several specific pollutants, such as ammonia, formaldehyde, phenol, and heavy metals (cadmium, copper, zinc, and lead), common components of pesticides or fertilizers, was below the maximum threshold, found on the page 27 of the report.

In general, the spray methods used to apply pesticides over large areas are significant contributors to airborne toxicity in farming. This method disperses high concentrations of chemicals as aerosols into the air. Some of these chemicals land on targeted crops, while others are carried away by the wind. As we know, air flows freely, moving at much higher speeds than other mediums that carry harmful chemicals. Unlike soil and water contamination, which can take weeks, months, or even years to spread to other regions via runoff or groundwater, air has no physical barriers and is the most efficient means of spreading pollutants globally.

Considering these factors, it is only natural that the high soil and water toxicity levels of the surveyed areas did not directly translate into the air quality results in the report. Instead, it highlights how the airborne toxicity generated by the country's farming activities has rapidly spread and contaminated other parts of the world.

Low and High Dosage Exposure: Moreover, an excerpt of another research paper which discusses the environmental and health impacts of harmful elements such as heavy metals raises a flag in this regard.

"Chronic low dosage exposure to numerous elements is a substantial threat to public health in many regions with metal pollution, particularly in places where metal pollution is ubiquitous."

https://www.sciencedirect.com/science/article/pii/S1018364722000465

A notable example is the trend in "Blood Lead Levels (BLL)" in people around the globe, which closely align with the periods of industrialization, increasing proportionally with the industrial outputs including the expansion of pesticide use in agriculture and peaking during the decades when leaded gasoline was used to enhance vehicle performance. Following the global ban on the leaded gasoline, the BLLs have declined, although they remain markedly higher than the pre-industrial levels. This highlights the persistent environmental toxicity from the accumulation of heavy metals and other harmful substances in the ecosystems and organisms, impacting the health of human and animal populations. More on the health impact from harmful chemicals will be presented in **Chapter V: Solutions**.

Organic Cotton: With the heightened awareness regarding the environmental harm of the conventional cotton farming, some farmers have turned to growing organic cotton. They often present various certificates to appeal the organic nature of their cotton to consumers. However, it is important to understand that this may not necessarily imply a total ban on harmful chemicals and no negative environmental impacts. As in the nature of many other certification programs, this practice may involve presenting only a limited scope to appeal to consumers while intentionally or mistakenly missing important perspectives of environmental harm in their certificates and farming processes. As this subject merits a dedicated discussion, I will present more details and perspectives in **Chapter IV: Misinformation Crisis**.

More specifically, the organic cotton cultivation may involve certain methods to inhibit the growth of bacteria and insects, such as bollworms, particularly harmful for cotton. In some cases, genetically modified cotton, such as Bt cotton, is used to reduce damage from insects. Similar to concerns over other genetically modified crops and vegetables, there is apprehension regarding the long-term effects of these modified crops on the ecosystems and human health. Furthermore, organic pesticides and fertilizers may be used instead of synthetic counterparts. However, despite the term "organic" providing reassurance to many, any substance designed to kill living organisms poses risks to human health and environment. This principle is similar to the comparison between organic and synthetic dyes, as organic dyes are not free from toxicity.

Moreover, the yield rate of the organic cotton can be lower, resulting in higher water and energy consumption, and consequently, higher quantities of greenhouse gas emissions. It may also require more agricultural land. Currently, organic cotton accounts for about 1% of the total cotton consumption globally, and this proportion is rising rapidly.

While I have some concerns about the organic cotton, I don't conclude that it is necessarily better or worse than their conventionally farmed counterparts. To be honest, I lack the comprehensive knowledge needed to fully understand its overall impact on the environmental sustainability. My point here is to emphasize that the true environmental impacts are rarely limited to a singular scope and require deeper analysis beyond the claims made by advocates. If a careful and thorough evaluation reveals a net positive effect, it should be supported as a way to reduce the significant environmental burden associated with cotton farming.

***

It is evident that cotton farming significantly influences the global environmental health. Therefore, I firmly believe that promoting natural fibers over synthetic fibers, solely based on a limited scope which favors associated arguments will not result in a healthier overall environment. Additionally, promoting synthetic fibers over natural fibers without considering complex factors such as consumers' tactile preferences, emotional connections, and the socioeconomic realities faced by natural fiber farmers worldwide would be equally counterproductive. Instead, efforts should focus on reducing the environmental impacts of both fiber types and making gradual, mid- to long-term transitions toward the most sustainable options.

10.1.2 Polymerization (Synthesis) of Synthetic Fibers

As discussed earlier, synthetic fibers are polymers created from the basic repeating units of monomers, derived from petroleum sources. The process of converting monomers to polymers is known as polymerization or synthesis. The overview of the polymerization and extrusion was discussed earlier in **Section 8. Synthetic fibers**. In this subsection, I present a commonly employed method for obtaining desired fibers from an intermediary material, known as polymer pellets.

Industry Specifics (14) – Start

Fiber Extrusion from Polymer Pellets

In some circumstances, the molten polymer in polymerization tank is converted into small, solidified pieces in the shapes and sizes similar with the grains of rice or barley – examples shown in **Chapter I: *Fig.22***, instead of extruding into the long strains of fibers. These small pieces are generally referred to as "polymer-pellets", "-granules" or "-chips".

Pelletizing and Conversion: As also reviewed in **Section 8. Synthetic fibers**, once the polymerization reaches a desired viscosity in the polymerization tank, the liquid undergoes the extrusion process that allows it to cool down and solidify. To produce pellets instead of fibers during this process, the polymer extrudes in larger thicknesses (similar to making thicker noodles in pasta machine), then cut to obtain the desired sizes of pellets. This process is generally referred to as "pelletizing".

In subsequent processes, the pellets are remelted in a melting tank and converted into various forms such as fibers, films, plastic molds, zippers, and more. For instance, a fiber manufacturing company can purchase these pellets, put them into the melting tank at their facility, and extrude fibers from the tank. Similarly, a zipper manufacturing company can melt them, and cast the molten polymer into the shape of plastic zippers. A water bottle company can follow a similar process to mold the liquid polymer into the shape of bottles. This method of using pellets to convert into desired shapes is akin to 3D printing, where plastic granules serve as source material, melted to make it flow, and then formed into specific shapes, thanks to the fluidity of the molten polymer and its solidification characteristics.

Rationale of Intermediary Pelletizing Process: Many may question the necessity of converting the liquid viscous polymer formed in the initial polymerization process into pellets only to reheat it in subsequent processes, considering the additional energy consumption this process entails. The rationale lies in the nature of the polymerization process itself.

Polymerization is an extensive and complex process requiring significant investment for large machinery setups and expansive land areas. Often, hundreds or even thousands of engineers, machine operators, and administrative staff work collaboratively in a single facility.

In the meantime, many factories producing fibers, zippers, bottles, and other plastic products lack the capacity to house such resource-intensive polymerization processes within their facilities. For instance, a zipper company with tens of employees only focuses on producing high-quality zippers. Such a company may not have necessary resources or interest to invest in extensive and costly polymerization processes. Instead, they can install a simpler, smaller, and less expensive melting tank to melt the pellets, then turn the molten polymer into desired shapes. This scenario is akin to a 3D printer owner who would not undertake the installation of a vast polymerization setup, even if he or she had a football field sized backyard and enough money. Consequently, the demand and market for various types of polymer pellets always exist and are substantial in size.

Melt-Blown Process

An illustrative example of the use of polymer pellets is a non-woven manufacturing process, specifically the "melt-blown" process. This process became critically integral to our daily lives when the COVID-19 pandemic broke out as it is used to produce face masks with high-filtration grades, such as N95 in US and FFP2 in Europe. Other examples include various goods consumed routinely by the public like sanitary pads, diapers, automotive filters, etc.

Significant Shortage and Efficient Industry Response: At the beginning of the pandemic, there was a significant shortage of face masks, especially those with high filtration grades, worldwide. Mask prices increased several folds within days and weeks, yet they were not accessible for weeks by the general public. However, the manufacturing capabilities of the melt-blown fabrics increased rapidly and responded to the public needs swiftly.

If the melt-blown manufacturing companies had to install complex and heavy setup of polymerization tanks to convert monomers into polymers, then into masks, building such facilities would have taken years. However, due to the extremely efficient and nimble industry supply chains, thanks in large part to the flexibility involving this pelletizing process, the global demand for the high-filtration masks was quickly met.

Industry Specifics (14) – End

Face Masks made with Natural Fibers: On a related note, masks made from natural fibers are significantly less efficient in blocking pathogens. As discussed earlier, natural fibers carry much higher moisture contents and it leads to creating a more favourable environment for the growth of microorganisms. Since human breath contains nutrients for microorganisms, masks made of natural fibers could easily become a habitat for various pathogens, such as viruses and bacteria. If humanity had relied on masks made from natural fibers, the pandemic would likely have caused much more damages. On the other hand, polypropylene used as the main ingredient of the high filtration masks is one of the most hydrophobic materials available, with a moisture content of less than 0.1%.

10.1.3 Polymerization (Synthesis) of Semi-Synthetic Fibers

As reviewed earlier in **Section 9**, the principles of producing semi-synthetic fibers are largely similar to those for synthetic fibers. Both involve polymerization in molten state to create a viscous liquid, which is then passed through spinnerets to solidify into thin, long fibers. The primary distinction lies in the source of the monomers: synthetic fibers are derived from petroleum resources, whereas semi-synthetic fibers originate from natural materials, such as wood pulp, using harmful solvents.

10.2 Yarn-Making

As many people may already know, yarn is an intermediary material necessary for fabric manufacturing. It is also utilized as sewing thread, ropes, and in various other applications outside fabric production.

Fiber Types: There are two primary types of fibers based on their form factors: staple fibers and filaments. Filaments are long, continuous fibers produced synthetically. In contrast, staple fibers are shorter and naturally found in materials like cotton and wool, or they can be created by cutting long synthetic fibers into short lengths.

Thick filaments can be used directly as yarn because they provide sufficient strength and necessary properties on their own for certain applications. A good example is the fishing line, which is a continuous synthetic fiber in its original extruded form. In other cases, thinner filaments are twisted together to achieve certain characteristics which cannot be obtained from a single strand of a thick filament.

Choosing fiber types and form factors like filaments or staples is a part of textile engineering, guided by the desired characteristics of finished goods such as texture, pliability, strength, and appearance, among others.

Spinning Principles: Staple fibers are too weak and short by themselves to be used for most textile applications except for the non-woven fabric manufacturing. To convert short and weak staple fibers into yarn, a process called "spinning" is employed. The spinning process arranges many strands of short fibers into a collection in length direction, multiplies them together, and finally twists the fiber group while adjusting the quantity of fibers in the finished yarn by pulling and winding onto a spinning stick. As a deeper exploration in the spinning processes extend beyond the scope of this book, I omit further details. If any reader is interested, the YouTube link below provides comprehensive overviews.

https://www.youtube.com/watch?v=YC638CmoOsU&list=PLkeQwaJxiD7ijEx-QkpX-YC4ye0xod8am

Dyeing: In some instances, yarns are dyed to achieve the desired look and feel of finished textiles before undergoing subsequent knitting or weaving process. If a yarn is not dyed prior to weaving or knitting, the resulting fabric is then dyed afterward to obtain specific aesthetic qualities. Additionally, dyeing can be performed on finished garments, particularly for highly customized and promotional items for specific events.

From a sustainability perspective, it is important to note that fibers and fabrics even in white color are typically dyed, contrary to common belief. The natural colors of natural fibers post-harvest are not pure white as they often contain various imperfections and impurities along with unique color tones formed in cultivation. For example, cotton has a yellowish tint while wool is in greyish shade post-harvest. To achieve a pure white appearance, various methods such as bleaching and dyeing are employed.

Given that dyeing imposes significant environmental burdens, it is crucial to highlight its impact on the environment. In contrast, virgin synthetic fibers present uniform white or transparent colors, while recycled fibers come with greyish tint due to the presence of contaminants, discussed earlier. Although the degree of a desired whiteness can determine whether a virgin synthetic material is dyed or not, the likelihood of synthetic materials not requiring dyeing process, thus benefiting the environment, is higher than that of natural fibers. This is apparent with widely used plastics such as bags, bottles, food containers, transparent films, etc., which do not typically contain any dyeing chemicals.

10.3 Fabric-Making

There are four primary types of fabrics commonly used in textile applications: woven, knitted, non-woven fabrics, and naturally grown sheet materials such as leather and fungi-grown mycelium textiles, also known as vegetable leather.

The desired characteristics of finished products determine the appropriate manufacturing methods of these fabrics.

Except for the leather, the other three fabric types offer a variety of particular construction techniques. For instance, woven fabrics can have visibly distinctive patterns such as plain, twill, satin, herringbone, jacquard, etc., on their surfaces. Knitted fabrics can include patterns like plain, double lock, warp knit, etc. In contrast, non-woven fabrics are constructed by randomly depositing fibers to form a fabric form factor, thus neither knitted nor woven. Therefore, it bypasses the yarn-making process and does not typically create any noticeable surface pattern.

Throughout human history, a vast array of fabrics has been developed, particularly in the woven and knitted construction methods. Given the extensive diversity, it is impractical to explain different fabric patterns comprehensively here, nor is it necessary for the environmental theme of the book. Therefore, I will provide the most basic concepts of the three main fabric types; woven, knitted and non-woven.

Woven: Warp yarns, running in length directions, and weft yarns in cross directions intersect to form a woven fabric structure. Patterns formed by crossing yarns determine fabric structures such as plain, twill, satin, etc. While traditional woven fabrics are known for their strength and non-stretch nature, there are also stretch woven fabrics where a portion of the comprising yarns impart stretch properties. Such yarn examples include spandex and elastane. Woven fabrics find widespread use in various applications including apparel, home textiles, outdoor gear, bags, etc.

Knitted: Knitted fabrics are formed by interlacing one strand or more into loop formations. Unlike woven textiles, knitted fabrics are inherently characterized by their stretch properties, which stem from these loop structures. However, various other knitting techniques creating different fabric constructions also exist that restrict structural movement, and a typical example is a warp knit. Examples of knitted fabric applications include sweaters, shirts, socks, bath towels, carpets, among others.

Non-Woven: Non-woven fabrics are created by intertwining numerous strands of fibers to form sheet structures. Unlike woven or knitted fabrics, the primary characteristic of the non-wovens fabrics lies in their low-density structure and capacity to hold large volumes of air. This attribute stems from the absence of the yarn-spinning process where fibers are bundled up and compacted by twisting. The voluminous structures of non-woven fabrics lend themselves well to specialized functionalities required in products such as face masks, sanitary pads, thermal insulation, filters, etc.

10.4 Dyeing

Dyeing process in textile manufacturing encompasses various form factors, including fibers, yarns, and fabrics. Analogously, in the leather industry, this process is known as "tanning". Moreover, dyeing may occur on finished goods following the final manufacturing stage of textiles, known as "cut-and-sew" process.

Chemical Intensiveness and Toxicity: As reviewed earlier, dyeing is almost always the most chemically intensive and toxic process of all textile manufacturing except for farming natural fibers where more toxic chemicals such as pesticides may be used. The chemical compositions of dyes and associated chemicals encompass a wide array of harmful elements and compounds, including heavy metals such as lead, chromium, and mercury, aromatic amines, halogens like chlorine and bromine, solvents, additives, chlorinated phenols, and others.

Significant Misperceptions: There is a common misconception that the toxicity associated with dyeing is limited to manufacturing facilities and does not affect consumers. Unfortunately, this belief is incorrect and contributes to a significant misunderstanding about the sustainability of textiles.

Many of us have likely noticed color fading in textiles after multiple home launderings. This is direct evidence of color pigments and other chemicals used in manufacturing being released into water, circulating through the water streams globally, and exposing us to the effects of these harmful substances in our daily lives.

Textile dyeing is far more complex than simply adding color pigments, as one might with food coloring in cooking. It involves a wide range of additional chemicals to ensure that dyes penetrate the fiber structures, either chemically or physically, and remain in the materials for as long as possible. Poor dye fixation, for example, can result in noticeable color loss and the transfer of the dyes to other textiles during home laundering.

While textile goods stored in a closet typically pose little risk, as the concentration of harmful chemicals is low in isolation, the situation changes dramatically when large amounts of textile waste are buried, burned, or left in dumping grounds without proper environmental control. In these cases, harmful chemicals become highly concentrated and are released into air, soil, and water through natural environmental processes such as wind, runoff water, and bacterial degradation. This can be observed in many regions where mountains of textile waste are destroyed with no environmental safeguards. A clearly visible example often seen in these circumstances is the slow leaching of heavily colorized pollutants from harmful chemicals in textile waste into river streams, resulting in ecosystem destruction, evident in the discoloration of water and the death of fish. Similar impacts occur in landfills and incineration sites, where pollutants spread through air circulation or contaminate groundwater.

10.4.1 Chemical Consumption and Environmental Impacts

As previously mentioned, dyes and associated chemicals contain harmful chemicals, including heavy metals and various compounds, each playing crucial roles in achieving specific colors and desired color-fading resistance. For instance, removing certain heavy metals from a particular dye formulation to reduce environmental toxicity can diminish the color tone and vibrance perceived by consumers.

To illustrate the diversity of chemicals employed in dyeing processes, it may be beneficial to review the range of substances supplied by a single chemical company - While most of the specific chemical names presented below may not carry much meaning to most readers, the long list serves the purpose to highlight the diversity and quantity of chemicals used in dyeing, contradicting a notion that simple color pigments are used.

- Basic chemical: Soda ash, Hydrochloric, Hydrogen Peroxide, Sulphuric acid, Acetic acid, Formic acid, Caustic soda, etc.

- Washing/Soaping agent: Serafast CRD, Kappatex R98, Seraperse CSN, Crosden LPD, Resotex WOP, Diypol XLF, Jintex WRN, etc.

- Detergent/Scouring agent: Jintex-GD, Felosan RGN, Jintex-GS, etc.

- Leveling agent: Levelex-P, Jinleve leve-RSPL, Serabid-IP, Dyapol XLF, Lubovin-RG-BD, etc.

- Salt: Common salt, Glauber salt, etc.

- Sequestering agent: Resotext 600S, Heptol-EMG, Heptol-DBL, etc.

- Whitening agent: Uvitex2B, Uvitex BHV, Bluton BBV, Tuboblanc col, Uvitex BAM, Synowhite, Hostalux ETBN, etc.

- Fixing agent: Sandifix EC, Tinofix-ECO, Protefix-DPE-568, Jinfix-SR, Optifix-EC

- Softner: Cetasaft CS, Resomine Supper, Acelon, Resosoft-XCL, Silicon, etc.

- Reducing agent: Hydrose

- Stablizer: Stablizer PSLT, Kappazon H53, STAB, Tinoclarite CBB, etc.

- Enzyme : Tinozyme 44L, Rzyme 1000, Avozyme CL PLUS, Enzyme-B50, etc.

- Anticreasing agent: Kappavon CL, Biovin 109, AC-200, Cibafuid-C, MFL, etc.

- Antifoaming agent: Jintex TPA, AV-NO, VO, Cibaflow-JET, etc.

- PH controller: Soda ash, Acid, Caustic, Neutracid RBT

I grew up in Taegu, South Korea, a city known as a major hub for textile dyeing during my childhood where the creeks and rivers in the city often resembled the images depicted in *Fig.36*.

Fig.36 Contaminated Water Streams

South Korea has undergone rapid development from one of the world's poorest countries following the Korean War to becoming an OECD member country today. This transformation has been accompanied by the establishment of stringent environmental regulations. Notably, visible pollution events such as contaminated rivers and creeks were swiftly addressed through strict controls on wastewater from industrial and residential sources. Consequently, dyeing factories were required to install water treatment facilities to meet government quality standards before discharge into natural water bodies.

However, while traveling to many underdeveloped and developing countries where environmental regulations may be less stringent, I still encounter numerous instances of water pollution of this nature.

As repeatedly mentioned, water circulates freely and globally through oceans, rivers, and precipitation, meaning that environmental impacts from distant locations can affect us no matter where we live within days, weeks, or months. Moreover, harmful chemicals persist in nature for extended periods, entering our ecosystems through food sources, air and drinking water, regardless of geographical location or dietary habits people have.

10.4.2 Dye Types and Toxicities

There is a wide variety of dye formulations used in manufacturing, chosen based on such variables as fiber types, specific colors, vibrancy, color fastness (fading resistance against washing, abrasion, etc.), and efficiency of dye absorption.

Dye Type Categorization: For the purposes of this discussion, I categorize the dyes into two groups: chemically reactive dyes and non-reactive dyes. Chemically reactive dyes, referred to as "Reactive Dyes", form bonds with the chemically active functional groups of fibers. Examples include the hydroxyl groups in the cellulose of cotton and the proteins in wool. In contrast, chemically non-reactive dyes are used for fibers lacking chemically active functional groups, such as polyester and polypropylene. A common type of dyes used for these fibers is called "Disperse Dyes". As the name indicates, this type of dye disperses and impregnates between the polymer chains of synthetic fibers by physical means rather than chemical bonding.

Color Fading: Understanding how different types of dyes interact with various fibers involves numerous scientific details. However, the basic concepts of the environmental impacts from different dye types can be grasped through simple observations in our own textiles. Many readers, including myself, may have observed that colors on textile goods made from natural fibers tend to fade more quickly in home laundering than those on synthetic fibers. A dark colored cotton shirt, for instance, can loose its color noticeably after several washes at home whereas a similarly colored polyester does not. This phenomenon is attributed to the chemically reactive nature of both the fibers themselves and a wide array of chemicals used in dyeing.

Chemical Reactions in Water: The chemically functional groups of natural fibers and reactive chemicals become activated in water. Water molecules exhibit polarity due to the presence of a partial positive charge on hydrogen atoms and a partial negative charge on oxygen, promoting chemical reactions with neighboring substances. Reactive dye pigments and other chemicals loosely fixed to natural fibers' functional groups interact easily with water and result in detaching themselves from the functional groups and forming bonds with neighboring water molecules. Color fading on textiles or discolored water during or after laundry are evidence of such interactions and subsequent pollution in water from the loosened chemicals. This phenomenon can lead to "accidents" in home laundry, where color migrations occur from one item to another during laundry.

In contrast, the disperse dyes and non-reactive chemicals used in most synthetic fibers do not create such chemical reactions in water, resulting in considerably lower pollution and risk of "accidents".

Organic Dye: It is important to address another common misconception surrounding "organic" dyes. With the growing environmental awareness, there has been a trend toward using naturally sourced dyes, also known as "organic" dyes, instead of synthetic equivalents. Many people assume that natural dyes, derived from plants, insects, and minerals, are free from negative environmental and health impacts. However, this belief is misguided.

While natural dyes do come from natural sources and typically do not contain harmful chemicals on their own, they have significant drawbacks, including producing duller colors and having poor adhesion to fibers, which leads to significant fading and the "washed out" effect during use. Combined with the heightened chemical reactivity of natural fibers, these issues pose challenges for textile applications. To address this, additional chemical agents known as "mordants" are often used to achieve desired tones, enhance vibrancy, and improve color fastness. Many of these mordants are metal-based and contain harmful elements such as chromium, copper, tin, and iron. In fact, the overall environmental impact from these natural dyes can be more severe than synthetic dye formulations due to the high toxicity level of mordants.

Despite the use of these toxic chemicals, textiles with natural dyes tend to lose their colors more easily during home washing than those using synthetic dyes. This often forces consumers to resort to drycleaning, which introduces further environmental harm.

10.4.3 Water Consumption

Textile manufacturing is widely recognized as the second largest consumer of freshwater globally, following the agricultural industry. While the agriculture primarily uses freshwater during farming (and sometimes for a final round of washing before consumption), textile goods require continuous freshwater use throughout its lifecycle. This includes excessive amounts of water consumption during the cultivation of natural fibers and laundering textiles over many years during consumer use.

When considering the total freshwater consumption across its full lifecycle encompassing numerous home washing cycles, textiles may even surpass the agriculture to become the largest consumer of freshwater resources – As there are large variations in both farming and laundry practices, it is difficult to compare the two accurately. Thus, this statement is based on my personal estimate.

Dyeing and Rinsing: During a typical dyeing process, textile materials such as yarns and fabrics are immersed in a large water bath containing a combination of necessary chemicals, including color pigments. These chemicals are then chemically activated for the reactive dyes and distributed for the disperse dyes in the water, facilitating the fixation of dyes and related chemicals into the chemical and physical structures of the fibers. The coloration process typically occurs within the same enclosed machinery system, ensuring all chemicals work uniformly to achieve desired colors in the fibers.

Following the coloration phase, multiple rounds of rinsing occur within the dyeing machines. Proper rinsing is crucial; inadequate rinsing can result in visibly adverse effects such as color fading or contamination onto other textiles during subsequent home launderings. In the initial rinsing stages, a significant amount of residual chemicals is removed, and the color of the water used in rinsing serves as an indicator of the extent to which the chemicals were not absorbed into the fibers during dyeing. Subsequent rinses involve multiple rounds of freshwater injections into the machine, progressively reducing the quantity of the excess chemicals until no color is detectable in the rinsed water.

Dyeing facilities implement various strategies to optimize chemical usage during the dyeing process, aiming to minimize the chemical quantities while maximizing effectiveness in achieving desired colors. However, in practice, excess quantities of chemicals are often injected to ensure successful coloration in the initial dyeing attempt. Failing to achieve the desired color in the first attempt can lead to significant challenges and costs in correction. To mitigate this risk, excess chemicals are applied to saturate the fibers to attain target colors, which, in turn, necessitate more water usage to rinse off the residual chemicals.

In some cases, "screen printing" techniques are employed in certain types of dyeing process, particularly for fabrics featuring intricate designs such as floral patterns. Unlike the water-immersion dyeing method explained earlier, the screen printing deposits dyes and related chemicals through screens, eliminating the use of water during coloration. However, thorough rinsing is still essential to remove excess dyes and chemicals after the printing.

***

As briefly mentioned in the earlier chapter, readers who dye their hair can easily understand some of the dyeing principals, particularly when dyeing wool fibers using the screen printing method. Like wool dyes, which are reactive and designed to chemically bond with the protein in wool, hair dyes also involve a chemical reaction with the protein contained in human hair. Hair dyeing doesn't require submersion in water, mimicking the screen printing method, but excess dye is necessary for a successful result, followed by thorough rinsing with significant water usage to remove the unabsorbed dye. Additionally, the dyed hair color fades over time relatively quickly, and proportionally with the frequency of washing. These characteristics closely resemble the process and behavior of natural fiber dyeing.

Room for Misunderstanding on Internet: There is a wealth of information available on the internet discussing the water consumption of textile materials and goods. However, many sources present data without citing specific references, leading to varying degrees of accuracy among the figures quoted. Some examples are listed below:

- 200 tons of fresh water per ton of dyed fabric -
  https://www.theconsciouschallenge.org/ecol

ogicalfootprintbibleoverview/water-clothing#:~:text=It's%20estimated%20that%20a%20single,per%20ton%20of%20dyed%20fabric

- About 1,800 gallons (approx., 6,800 liters) of water are needed to produce a pair of jeans – https://www.oldhamcountywater.com/interesting-water-facts.html#:~:text=About%201%2C800%20gallons%20of%20water,about%20150%20gallons%20of%20water

- 75 liters of water goes in making your jeans (YOUTUBE) – https://www.youtube.com/watch?v=3PKzGvSaWYA

- It takes 2,700 litres of water to make one cotton t-shirt. That's enough water for one person to drink for 900 days or almost two and a half years! – World Wildlife Fund (WWF), https://www.linkedin.com/pulse/how-much-water-does-take-make-t-shirt-md-jubair-hasan/

With the lack of scientific evidence, it is challenging to determine the reliability and accuracy of such information. For instance, one source claims that producing a pair of jeans requires 6,800 liters (1,800 gallons) of water, while another states only 75 liters (20 gallons), resulting in a discrepancy of approximately 90 times.

Clearly, water consumption can vary widely based on many factors such as agricultural practices, weather conditions specific to individual cotton farms, dyeing methods, etc. Given these variations, it is crucial to approach such information with caution.

Lack of Balance: Moreover, a significant issue with these types of data is the frequent omission of the differences in water consumption between different materials, for example, cotton and polyester, as this omission may lead to a considerable misunderstanding on the environmental impacts of different fiber types.

For instance, the first bullet point, stating 200 tons of fresh water per ton of dyed fabric, lacks the disclosure of a significant difference in water consumption between natural and synthetic materials. Typically, dyeing natural fibers incurs more rounds of rinsing cycles, requiring more amount of fresh water, compared with synthetic fibers. However, the figure of 200 tons does not address such discrepancy.

One example specifying water consumption by a material type is the bullet point citing World Wildlife Fund (WWF), which states that 2,700 liters of water are needed to produce a cotton t-shirt. While this information is more precise, it would have been more helpful to compare it with the water consumption of an equivalent polyester t-shirt. While I understand that the purpose of the WWF article might not be to provide scientific comparison for different textile materials, the lack of opportunities for the public to gain more comprehensive understanding contributes to the widespread misinformation and misperception crisis we face today – more discussions in **Chapter V: Misinformation Crisis**.

10.5 Cut-and-Sew

In textile manufacturing, the "cut-and-sew" process is typically the final stage before the finished goods are packaged and transported to various retail fronts. This process requires minimal investment and can be operated with a few sewing machines, skilled workers, and a modest space to manage the operations. It is akin to setting up a workspace in a corner of the living room in a household with a sewing machine. Numerous manufacturing facilities around the world engage in these cut-and-sew operations.

Waste from Cut-and-Sew: Large-width fabrics, commonly 152cm (60 inches), are cut into the shapes and sizes of the finished goods, which are then sewn together to create the final products for consumption. A crucial aspect of this process is the significant amount of waste it generates. Depending on the shapes and sizes of finished goods, this cutting process can easily produce 5 ~ 30% waste from the incoming fabrics. An example is shown in *Fig.37* where the blue part (1) will be sewn together to make a shirt, and the yellow part (2) will be unwanted after the cutting.

Fig.37 Waste (2 in yellow) after Cutting

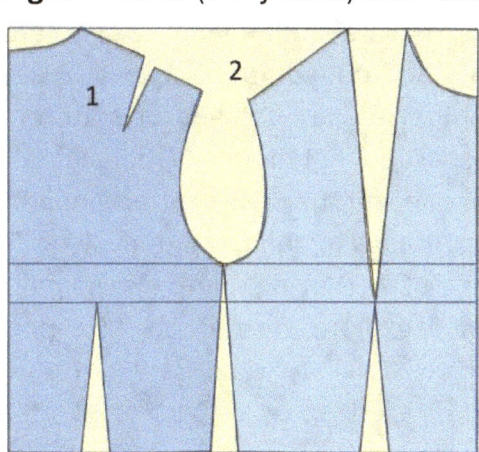

While some factories may sell the leftover materials to manufacturers who use them as stuffing for, for example, teddy bears, such applications are quite limited. As a result, most factories treat these leftovers as industrial waste, paying for either burial or incineration. Considering that tens of billions of textile products are produced annually, the scale of this waste and its environmental impacts are massive. Moreover, the amount and chemical reactivity of reactive dyes and chemicals lead to consequential environmental impacts from enormous quantities of waste from the cut-and-sew process.

Given the importance of effectively handling the waste from this process, an in-depth analysis and solutions will be discussed in **Chapter V: Solutions, Section 21. Clean Recycling Initiative™**.

10.6 Other

Subsequent to the cut-and-sew, typical processes include packaging, transportation, stocking, and retailing. Some of these processes have sparked debates, such as the amount of plastics used for packaging, methods of transportation, etc.

These concerns are independent of material types, manufacturing and consumption cycles in environmental aspects. Whether a shirt is made of cotton or polyester, these processes do not create a considerably different impact from one another. Given the simple nature and straight forward assessment of the involved factors such as the quantity of packaging materials, shipping methods, etc., I do not explore further on this part of textile manufacturing.

Chapter II Post-Chapter Commentary

In this chapter, I covered a broad scope of textile materials and manufacturing processes with two primary objectives. The first is to provide readers with a basic understanding of textile goods from their origin materials to manufacturing and to packaging before consumption. The second is to cultivate analytical mindsets and skills on related subjects, enabling them to identify significant gaps in numerous environmental claims and actions made by interest groups and activists.

As a disclaimer, I do not dismiss the concerns about the negative impacts of synthetic fibers. Shared earlier in several parts of this book was my professional role as textile engineer in South Korea, where I produced synthetic fibers with my own hands and participated in nearly all stages of manufacturing processes. This experience allowed me to gain hands-on knowledge and comprehensive views on different textile materials and manufacturing processes. From this, I have been raising awareness and vocal about the health impacts of microplastics for many years by now.

I reiterate that I do not deny the negative impacts of synthetic fibers. However, we must acknowledge that we do not live in a utopian world where all problems vanish by replacing one problematic material with another. My point throughout the book is that we must consider and evaluate the true overall impacts with comprehensive views and throughout the entire lifecycles of the materials in question. Favoring one material type over another based on limited scopes may result in creating more detrimental effects on the environment.

In the following chapter, readers will gain more insights based on the evaluation criteria I developed, consisting of sixteen separate segments for the Lifecycle Assessment (LCA) of cotton and polyester.

Chapter III

***

Lifecycle Assessment (LCA)

Pre-Chapter Commentary

Quote of the chapter

"Many of these factors represent important dimensions of our lives, that cannot be ignored nor easily altered. While I have had a firm belief in the environmental benefits of synthetic fibers over natural ones and have reflected it in my purchasing decisions wherever applicable, I do not expect such conviction will transfer to others, significantly changing their shopping patterns in favor of synthetic materials, nor should they, as there may be too many complex socioeconomic implications."

***

In this chapter, I will adopt a more systematic approach to examining the environmental impacts across the entire lifecycle of two material categories: natural and synthetic fibers. The goal is to provide a science-based and comprehensive perspective, moving beyond the commonly held, notion-driven misjudgements and limited evaluation parameters.

While many Lifecycle Assessment (LCA) reports from various environmental groups tend to focus primarily on carbon footprints and their implications for climate change, this chapter will take a broader approach.

Overview

The LCA analysis method I developed includes the sixteen evaluation criteria illustrated in **Fig.38**. While the carbon footprint, highlighted in the map, is often discussed in the context of petroleum-based materials, it is also closely linked to the farming of natural fibers, as agricultural activities require significant energy, either directly or indirectly, for cultivation. In fact, many readers might be surprised to learn that some natural fibers generate a considerably higher carbon footprint than synthetic fibers.

Furthermore, it is not only the carbon footprint perspective from the source materials important but also various other critical evaluations such has toxicity, water usage, food impact, etc., are critical to assess the overall impacts of different materials. This analysis is designed to cover the environmental sustainability scope as comprehensively and objectively as possible.

Fig.38 LCA Evaluation Criteria Map*

Evaluation	Farming/Fiber making	Manufacturing	Consumption	Disposal
Toxicity Footprint Index (TFI)	1	2	3	4
Water Footprint Index (WFI)	5	6	7	8
Food Footprint Index (FFI)	9	10	11	12
Carbon Footprint Index (CFI)	13	14	15	16

* The numbers and shading on the map do not reflect environmental impacts or significance.

Before diving deeper into the analysis of each evaluation criterion, I believe it is important to first share my perspectives in approaching a wide range of environmental concerns and establishing necessary actions to address them more effectively:

Climate Change Concerns

It is undeniable that global warming has significantly impacted our lives. While many climate-change models predict detrimental consequences in many different angles, I refrain from quoting specific predictions and numbers here as we can easily see this through a wide range of scientific data, reports and books readily available to the public. For instance, I follow the World Meteorological Organization (WMO) on social media and find that their work of presenting relevant information visually and impactfully is impressive. Faced with an abundance of data suggesting that we are falling short of carbon and temperature goals, one cannot help but feel deeply concerned by significant challenges ahead.

Weather Impacts: As alarming as it sounds, I approach these challenges from a slightly different perspective. For instance, when examining the top 10 largest natural disasters by death toll and comparing them to the global temperature changes, as depicted in **Fig.39** through **44**, it becomes evident that these worst events are not linked to the global warming we are experiencing of late. Many of these disasters are earthquakes, or they occurred before the cumulative carbon effects began influencing the global temperatures.

Fig.39 Top 10 Natural Disasters by Death Toll

	Death toll (Highest estimate)	Event	Location	Date
1	4,000,000[1][a]	1931 China floods		July 1931
2	2,000,000[2][3][4]	1887 Yellow River flood	China	September 1887
3	655,000[5]	1976 Tangshan earthquake		July 28, 1976
4	500,000[6][1]	1970 Bhola cyclone	East Pakistan (now Bangladesh)	November 13, 1970
5	316,000[7]	2010 Haiti earthquake	Haiti	January 12, 2010
6	300,000[8]	526 Antioch earthquake	Byzantine Empire (now Hatay/Turkey)	May 526
7	≈300,000[9]	1839 Coringa cyclone	Andhra Pradesh, India	November 25, 1839
8	≈300,000[10]	1737 Calcutta cyclone	Bengal, India	October 1737
9	≈300,000[11]	1139 Ganja earthquake	Seljuk Empire (present-day Azerbaijan)	September 30, 1139
10	273,407[12]	1920 Haiyuan earthquake	China	December 16, 1920

https://en.wikipedia.org/wiki/List_of_natural_disasters_by_death_toll

In the scientific community, earthquakes, which account for half of these top 10 disasters, are not considered related to climate change. Additionally, all the cyclones and floods on the list took place before the industrialization significantly impacted global warming, a point that merits further investigation in the following.

While there may be different ways to interpret the information in **Fig.40**, which shows the global temperature changes over the last 500 million years, one conclusion is clear: warming and cooling are natural cycles in Earth's ecosystems.

227

Fig.40 Estimated Global Temperature over the Last 500 Million Years

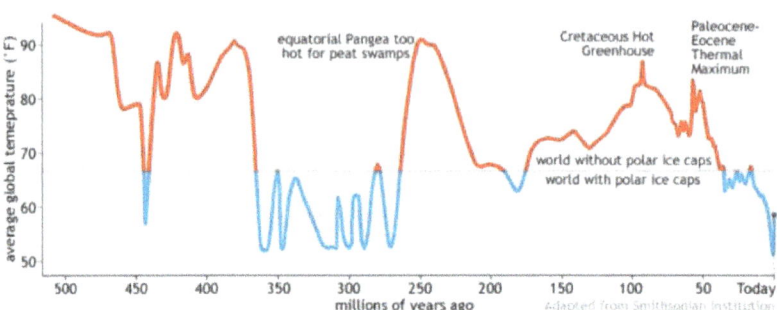

https://www.climate.gov/news-features/climate-qa/whats-hottest-earths-ever-been

Zooming in on more recent decades, as shown in **Fig.41**, the most recent temperature rise appears to have started in the late 1970s and early 1980s.

Fig.41 Global Average Surface Temperature

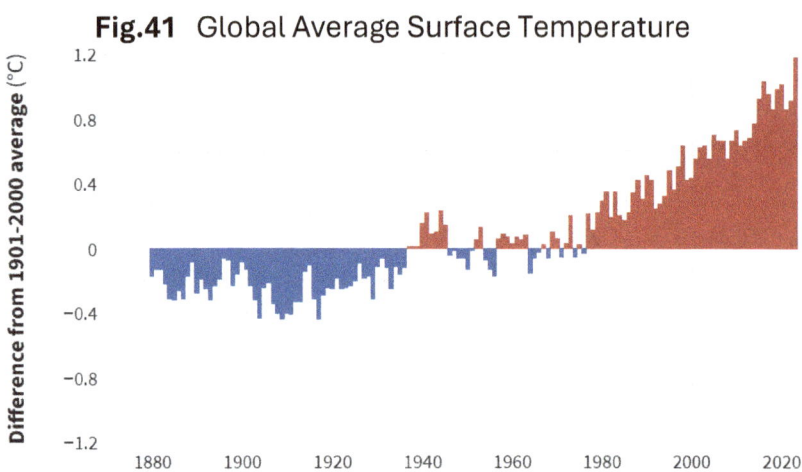

https://www.climate.gov/news-features/understanding-climate/climate-change-global-temperature

228

From a scientific perspective, it would be extremely challenging to demonstrate that the cumulative effects of carbon emissions led to a significant rise in extreme weather events immediately following the onset of the new warming trend in the late 1970s and early 1980s. A more rigorous approach would be to consider a "grace period," during which human-induced carbon impacts may have overlapped with natural and cyclical temperature changes. This grace period acknowledges that the initial phase of the most recent warming trend might not be fully attributable to human activities. Based on this understanding, none of the top 10 disasters can be linked to the ongoing global warming.

Expanding the scope to the "Top 50 Disasters by Death Toll" reveals a similar pattern. Although the top 50 list is not provided here, those interested can easily search for it using specific keywords. Notably, only one event, Cyclone Nargis in Myanmar in 2008, which ranks 15th, falls within the period where the cumulative carbon impacts might have had a significant influence. The rest of the top 50 list is predominantly composed of either earthquakes or weather-related disasters that had occurred well before any controversial periods associated with the modern climate change debates.

Another important consideration is that the modern infrastructure is far better equipped to handle cyclones and floods than in the past. Consequently, similar events today would likely result in far fewer casualties compared to historical disasters like the 1931 China floods or the 1887 Yellow River flood, in China which claimed 4 million and 3 million lives, respectively. This observation highlights humanity's growing ability to adapt and respond to changing environmental conditions, a theme that will be further explored in subsequent discussions.

More Pressing Threats: In my view, the more pressing threat from the climate change lies in its consequences, which can cause more immediate and abrupt impacts. For instance, the scientific understanding of the potential reappearance of ancient microorganisms, such as viruses and bacteria, from the thawing permafrost at the northern and southern extremities of the globe occupies my mind more profoundly, albeit none of the negative climate impact is desirable, because of the "toxic", "violent" and "invisible" nature of the impacts. A notable example occurred in 2016 when an anthrax outbreak in Siberia was linked to a thawing reindeer carcass buried in the permafrost of the region, resulting in the death of one child and the infection of dozens of people.

The concern is that the rapid melting of the permafrost in regions like Siberia and Alaska could release ancient pathogens that modern humans and animals have never encountered and against which they have no immunity. The most recent thawing permafrost trend has been observed since the 1950s, initially raising concerns due to its impact on various infrastructures in some northern regions. By the 1990s, research indicated that the permafrost was thawing at alarming rates, and we can expect this thawing to accelerate, potentially leading to higher impacts from the release of ancient pathogens.

While this area of concern is real, as evidenced by the 2016 Siberian case, it remains an ongoing area of research with many factors to consider. In my opinion, this issue warrants immediate attention and continued study, but without causing panic. Throughout history, humans have faced numerous pathogenic threats to which we had no prior immunity, or the pathogens became more potent to overcome any existing immunity. While some outcomes have been more severe than others, the potential risks from the pathogens emerging from the permafrost are not fundamentally different from other invisible threats we live with daily. This point naturally leads to the next topic of discussion.

Modern Way of Living: This perspective of mine also reflects the modern way of living. Advances in science have made us ever more knowledgeable than our ancestors, enabling us to recognize threats more quickly and effectively. Moreover, we live in an increasingly densified world, with rapid population growth leading to more concentrated systems, such as highly industrialized animal farming, which is believed to have contributed to the emergence and spread of the avian flu among many others. Population density likely also played a role in the rapid spread of recent pandemics like the COVID-19.

From another standpoint, humans have always faced risks. Without the benefit of air conditioning, heating, and sturdy homes, even relatively mild weather events by today's standards would have been far more concerning in ancient times. Looking back on my life, I can recall that weather was always a major topic between my parents and relatives in family gatherings, and this collective focus on it has always been a part of our society. Even in ancient times, it seems implausible that our ancestors failed to recognize such environmental threats. As such, we can easily understand that environmental dangers have always been a part of human existence. The difference today is that, thanks to scientific advancements, we are far more knowledgeable of these threats and have access to more precise information than ever before.

Need for Comprehensive Analysis: While I do not downplay the weather impacts of the climate changes such as rising sea levels, increased frequencies of severe weather, etc., I must emphasize that the causes characterized earlier by "toxic", "violent" and "invisible" nature cannot be overlooked.

Increasing toxicity of chemicals in the environment, as repeatedly discussed in the book, could soon become the most threatening factor for the humanity. Focusing solely or primarily on the carbon emission is akin to worrying only about energy consumed in cooking while neglecting to wash off the pesticides used in farming the ingredients. In reality, people pay attention to how the ingredients are farmed because of the recognized danger of the toxicity from the food sources, leading to thoroughly washing them before consumption, while the growing popularity of biological and organic ingredients has been a definite trend. Unfortunately, this principle is largely ignored in the areas of the global environmental and sustainability management.

Evaluation Criteria

The lengthy prelude of this chapter aims to justify the need for a broader and more comprehensive perspective on the environmental management, extending beyond just carbon impacts and climate change. Accordingly, this chapter breaks down the lifecycles of two key fiber categories - cotton, representing the natural fibers, and polyester, representing the synthetic fibers - into the following four segments: 1. Farming (for natural fibers) / Fiber Production (for synthetic fibers), 2. Manufacturing, 3. Consumption, and 4. Disposal. Then, the assessments are conducted in the following four evaluation criteria: 1. Toxicity Footprint Index (TFI), 2. Water Footprint Index (WFI), 3. Food Footprint Index (FFI) and 4. Carbon Footprint Index (CFI). Finally, Total Impact Index (TII) is calculated and presented at the end of the chapter.

Simplified Method of Index Analysis

In each area of the evaluation, I use a scoring system of "Low", "Medium", and "High". While this approach may not offer the precision of numerical scoring, it is crucial to understand that scientifically measuring and quantifying environmental impacts with absolute accuracy is inherently complex and challenging. Although some organizations or individuals may publish specific figures for certain environmental assessments, these evaluations often have limited scopes and focus on specific aspects.

For instance, the health impact of a toxic chemical can vary widely depending on numerous factors, such as exposure level and duration, an individual's age, gender, existing health conditions, and more. As a result, it is impossible to accurately quantify the impact of a particular chemical or group of chemicals on the human body in a way that universally applies to the entire global population. Similarly, the environmental impacts of textile materials are influenced by numerous factors. Given this complexity, objectively quantifying the full range of environmental impacts with numerical precision across various evaluation criteria is virtually impossible.

For these reasons, I adopt a simplified three-step scoring system of "Low," "Medium," and "High". I believe this method will still provide readers with a comprehensive understanding and balanced perspective, offering a reasonably accurate grasp of the overall environmental impacts of different fiber categories.

Focus on Cotton and Polyester

Given sufficient knowledge on the lifecycle of a specific material type, the analysis methodology I present here can be applied to any type of fiber. However, in this book, I focus on cotton and polyester. While numerous other types of fibers are used in textiles, I chose to analyze the two to maintain a focus on the most impactful and significant fiber types.

Throughout this chapter, I will also make special mentions of certain other fiber types when there are significant environmental factors to consider.

***

Finally, I acknowledge that the scores obtained in this analysis for each evaluation area may change in the future depending on advancements in various factors such as farming techniques, material technologies, and more. However, I do not anticipate any major shifts from the conclusions, as the fundamentals of natural and synthetic fibers will always remain consistent. If, in the future, disruptive technologies such as cotton grown in laboratories become a reality, I will make another attempt to evaluate them.

Section 11 LCA - Farming / Fiber-Making

This section examines the initial stages of the fiber lifecycle: farming for natural fibers and synthesis for synthetic and semi-synthetic fibers.

11.1 Toxicity Footprint Index (TFI) Analysis - Farming / Fiber-Making

Before delving into the toxicity analysis in this section, it is important to revisit a discussion from the previous chapter regarding the definition of "toxicity". While many environmental interest groups and activists often label synthetic fibers as "toxic", this characterization does not align with the traditional understanding of toxicity. Synthetic fibers exhibit minimal chemical reactivity, if any. When a material does not induce chemical reactions, it cannot accurately be described as "toxic". In fact, synthetic fibers are often referred to as chemically "inert", which is fundamentally the opposite of "toxic"

This distinction is critically important when evaluating the short-, medium-, and long-term impacts of these materials, and when strategizing effective actions to address them with prioritization on more imminent threats.

11.1.1 Cotton Cultivation

A significant majority of cotton fibers are cultivated with the use of pesticides and fertilizers. As discussed in the previous chapter, both pesticides and fertilizers contain highly toxic elements and compounds. For instance, pesticides are designed to eliminate living organisms such as insects, bacteria, fungi, actinomycetes, algae, etc., that negatively impact the growth of desired plants. Even without specialized knowledge in chemistry or health science, one can easily understand that pesticides are also harmful to other living organisms, including humans.

Pesticides

Pesticides have been utilized for thousands of years, with the earliest records dating back to 2500 B.C. The Sumerians used natural sulfur compounds, known for their strong odor, to repel bugs and insects. By 400 B.C., the Persians were using a plant-derived organic compound sourced from flowering Chrysanthemum plants to repel insects.

In the meantime, the use of naturally occurring ingredients in pesticides does not imply that they are harmless to the environment. When harmful substances are highly concentrated, they lead to environmental toxicity. A parallel can be drawn with mining, where the extraction of natural elements like copper, iron, gold, silver, etc., from the Earth crust results in the concentration of toxic elements, creating significant environmental toxicity. This issue will be explored in greater depth in **Chapter V: Solutions**.

Industrialization of Synthetic Pesticides: The history of pesticides took a significant turn in the 1940s, post-World War II, with the development of synthetic pesticides. These synthetic pesticides proved to be significantly more effective than their natural counterparts in creating more favourable outcomes in farming. For instance, DDT (dichloro-diphenyl-trichloroethane) was widely used not only for growing crops but also on humans and animals to repel mosquitos and other insects for decades - It was among the first synthetic pesticides, followed by a series of other toxic chemical compounds. Although DDT was banned in most countries between the 1970s and 1990s due to its potent impacts, earning it the nickname "silent killer", it is still used in some regions.

Toxicity and Persistence: A short description of the USA Environmental Protection Agency (EPA) to describe the nature of the chemical is as follows:

DDT is:

- known to be very persistent in the environment,
- will accumulate in fatty tissues, and
- can travel long distances in the upper atmosphere.

Some studies have shown that DDT can remain in soil for 15 years and can easily endure longer depending on soil types, temperatures, bacterial compositions of each soil type. Furthermore, its breakdown compounds, DDE (Dichlorodiphenyldichloroethylene) and DDD (Dichlorodiphenyldichloroethane), also persist in the environment and can be found for many years after the original application and breakdown of DDT.

Consequently, DDT and related chemical compounds can be found in the environment today, including in the bloodstreams, tissues, organs, breast milk, etc., of both humans and animals even in remote locations from their original sources.

Dirtiest Crop: Cotton is often referred to as the world's dirtiest crop due to its extensive use of insecticides, accounting for 16% of the global insecticide consumption. Various pests, such as bollworms, plant bugs, stink bugs, etc., are particularly attracted to cotton, necessitating the use of significant quantities of toxic agrochemicals to manage these pests. Given that cotton occupies only 2.5% of agricultural land, the disproportionately high percentage of the insecticide use highlights the considerable toxicity impact associated with the cotton cultivation.

Fertilizers

Regarding the toxicity of fertilizers, they are generally considered less toxic than pesticides, as the fertilizers are designed to promote plant growth rather than eliminate pests. However, this does not necessarily mean that what benefits plants is equally beneficial for humans. In fact, both organic and synthetic fertilisers create significant amount of greenhouse gas emissions from the release of CO_2, CH_4, NO_2, etc., while involving many other chemical impacts.

History of Natural Fertilizers: Similar to pesticides, fertilizers have a long history of use by humans. They were among the most accessible natural materials because human- and animal-waste, and other organic matters such as food waste, animal carcasses, etc., are rich sources of plant nutrients with such elements as nitrogen, phosphate, potassium, etc., generated from the natural decomposition of such substances. This long-standing relationship indicates that fertilizers have been utilized since the beginning of the agricultural history.

One of the earliest commercialized natural fertilizers was guano, the natural deposit of bird remains and droppings. However, like many other natural resources exploited the post-industrial revolution era, the guano quickly became scarce and led to conflicts over national interests and a source of imperialism in the early 19th century. For instance, United States passed the Guano Islands Act in 1856 to claim the ownership of the guano in the Pacific Ocean.

Industrialization of Synthetic Fertilizers: The fertilizer industry underwent a significant transformation in 1913 with the invention of the Haber-Bosch process by two German scientists Fritz Haber and Carl Bosch. This process enabled the synthesis of ammonia from atmospheric nitrogen in industrial settings at low costs, leading to the widespread use of the synthetic fertilizers and largely replacing the guano and other natural fertilizers. The Haber-Bosch process is recognized as one of the most important developments in the modern human history, as it significantly increased agricultural productivity and supported the rapidly growing global population. However, as with all other human acts, there were unintended side effects.

Densification and Concentration: Harmful elements and substances such as nitrogen, phosphate, potassium, etc., which are either initially contained in the fertilizers or produced as the byproducts of their use in agriculture, have led to widespread environmental issues. These elements are carried by runoff water and wind, contributing to the pollution in various water bodies and air streams globally. The core issue, whether the source is natural or synthetic, lies in the concentration and densification of these harmful elements. In their natural, dispersed states, these elements pose minimal risk to the ecosystems. However, when they become highly concentrated, they can cause significant environmental and health hazards.

***

Toxicity from farming natural fibers is easily the most harmful component of all environmental factors in textile manufacturing for most consumer goods. While dyeing is generally considered a chemically intensive process in the manufacturing segment, its toxicity level is incomparably lower than that created by the farming. Given these considerations, the Toxicity Footprint Index (TFI) of cotton in the farming segment is rated as "High" gaining the score of 3 in the three-step scoring system of this LCA analysis.

11.1.2 Polyester Synthesis

Polyester's journey to become consumable products begins with the synthesis of monomers into polymers as reviewed in the earlier chapter.

IARC Classification: To understand the nature of the chemicals involved in the synthesis of polyester, I use the classifications of International Agency for Research on Cancer (IARC) for estimated health impacts, particularly cancer, of different substances with the examples below – The IARC classifications are also used for broader health impacts beyond cancer:

Group 1: *"Sufficient evidence of carcinogenicity"* in humans, and these agents have been linked to cancer through epidemiological studies. Exposure can lead to respiratory issues, skin irritation, and sensitization. Examples - Tobacco and Tobacco Smoke, Alcoholic Beverages, Processed Meats, Diesel Engine Exhaust, etc.

Group 2B: *"Possibly carcinogenic to humans"*. There is some evidence that it can cause cancer in humans but at present it is far from conclusive. Exposure can cause irritation of the eyes, skin, and respiratory tract. Examples – Roasting fumes of coffee, some pickled vegetables, gasoline engine exhaust, etc.

Group 3: *"Not classifiable as to their carcinogenicity to humans"*. This classification is used when there is insufficient evidence of carcinogenicity in humans and inadequate or limited evidence in experimental animals. Essentially, it means that current studies do not provide enough data to determine whether these agents cause cancer. Examples – Caffein, chlorinated drinking water, fluorescent lighting, etc.

Chemicals for Polyester Synthesis: The key chemicals involved in the synthesis of polyester include Ethylene Glycol, which is classified as Group 3 by IARC, indicating it is not classifiable as to its carcinogenicity to humans. Similarly, Terephthalic Acid (TPA) also falls under Group 3, while Dimethyl Terephthalate (DMT) has not been classified by IARC, suggesting no significant health concerns associated with it. On the other hand, Antimony Trioxide and Acetaldehyde are classified as Group 2B, meaning they are "possibly" carcinogenic to humans, similar with such mundane components of our daily lives as roasting coffee fume, some pickled vegetables, etc., while Formaldehyde is classified as Group 1, known to be carcinogenic to humans.

Despite the use of Formaldehyde in polyester synthesis that fall into the Group 1, it is important to note that these chemicals are used invariably in closed manufacturing systems under strict controls, which significantly reduces or eliminates its exposure risk to workers. This contrasts sharply with the cultivation of natural fibers like cotton, where toxic chemicals such as pesticides and fertilizers are applied in open fields without recovery systems, leading to significant health risks for farmers who are exposed to these substances during and after the initial application.

Numerous deaths have been linked to the toxicity of pesticides and other chemicals used in cotton cultivation, as farmers are exposed to these substances through aerosols during spraying and residual contact during cultivation. In contrast, there have been no reported deaths linked to the exposure to Formaldehyde in the synthetic fiber production.

Operational Efficiency: Profit-driven organizations in the petroleum and plastics industries have become remarkably efficient at maximizing profitability while minimizing their environmental footprint, particularly in terms of their resource utilization and chemical management. These industries are among the leaders in efficient conversion processes, transforming raw materials into finished products with minimal waste, an achievement made possible by continuous innovation over their long histories. For example, almost every ounce of crude oil is refined into various fractions, and plastics manufacturing similarly achieves extremely high efficiency in material use.

Toxic chemicals involved in these processes are often reclaimed and reused, or they undergo strict environmental treatment. This is not only essential for their environmental compliance but also for the cost management, as accidental releases of harmful substances could result in significant financial losses.

Center of Attention: Oil refinery and polymerization facilities are typically located in industrial parks, often near major city centers to optimize transportation and supply chain efficiencies. These facilities are invariably subject to strict environmental regulations and are heavily monitored to ensure compliance.

During my tenure as engineer in South Korea, I worked in the facility, which included a large polymerization process for acrylic fibers, situated just 10 minutes from the core city center of Masan, a city of 700,000 ~ 800,000 residents at the time. The facility was surrounded by numerous residential houses and shared a river stream with its neighbors. It was therefore closely monitored to ensure its operations adhering to both local and federal environmental standards. While accidental releases of harmful chemicals can occur in any industrial settings, such incidents are extremely rare and, if it happens, it attracts significant public and regulatory scrutiny.

Later in my career, I served as an environmental representative at a manufacturing facility in Edmonton, Alberta, Canada, which produced various paper and roofing products. Any wastewater leakage into natural water bodies had to be immediately reported to the provincial government of Alberta, with potentially significant fines depending on the nature of leakage.

***

Due to the stringent monitoring and regulatory systems in place, it is rare for harmful chemicals to be released into the environment during polymer synthesis. This is in stark contrast to the virtually uncontrolled application of toxic pesticides and other chemicals by cotton farmers worldwide. As a result, the Toxicity Footprint Index (TFI) for polyester in the Fiber-Making segment is rated as "Low", with the score of 1.

The evaluation results for the Toxicity Footprint Index (TFI) of cotton, rated as "High", and polyester, rated as "Low", will be tabulated and presented at the end of this subsection in a graphical format alongside three other evaluation criteria in the Farming and Fiber-Making segment: Water Footprint Index (WFI), Food Footprint Index (FFI), and Carbon Footprint Index (CFI).

11.2 Water Footprint Index (WFI) Analysis - Farming / Fiber-Making

Water consumption is a crucial factor often overlooked in the sustainability assessments of both natural and synthetic fibers. To address this, I reference a research paper by Jennifer Xiaopei Wu and Li Li, which provides an in-depth analysis of overall water consumption for various fiber types.

As illustrated in **Fig.42**, over 90% of the water footprint is attributed to the natural and semi-natural fibers (the sum of cotton, wool, silk, flax/linen and viscose). Specifically, cotton and viscose together account for 81% of the total water footprint, with cotton alone contributing 60% and viscose 21%. In contrast, the synthetic fibers are responsible for only about 8% of the total water footprint across all fibers.

Fig.42 Water Footprint - Total Processing – UK, 2016

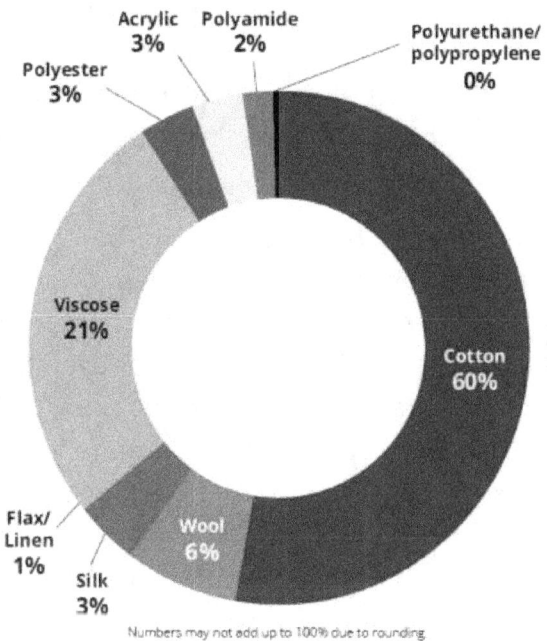

https://www.researchgate.net/figure/Water-footprint-for-the-total-processing-phase-of-each-fiber-type-for-the-UK-m3-in_fig3_335139274

When combined with the analysis of the total fiber consumption presented in *Fig.1* and *2* of **Chapter I**, it becomes clear that the synthetic fibers, which constitute 65% of the total fiber consumption, account for only 8% of the total water consumption, highlighting a significant advantage of the synthetic fibers. In contrast, the natural and semi-natural fibers, which make up 35% of the total fiber consumption, are responsible for 92% of the water footprint.

This stark difference arises from the nearly waterless polymerization process used in the synthetic fiber production, compared to a vast amount of water required for the farming. While this study examines the full lifecycle of the fibers and is not limited to just farming and synthesis, the contrast would be even more pronounced if focused solely on this segment, given the minimal water usage in the processing of most synthetic fibers.

The extensive use of groundwater by the agricultural industry has raised significant concerns about long-term sustainability. Given that groundwater replenishment is a slow process, often taking hundreds or even thousands of years, excessive extraction is exacerbating global water scarcity. Cotton, in particular, has a notorious reputation as "thirsty crop" for requiring more water than most other crops. Moreover, this high water consumption extends throughout its entire lifecycle, from farming and manufacturing to home laundering, further intensifying its environmental impact.

* A recent study published in 2023 found that excessive groundwater extraction has shifted Earth's axis of rotation by about 4.36 cm per year, totaling an 80 cm shift between 1993 and 2010. While I always take cautionary approach on the information of this nature and this finding has not been linked to any considerable harm in our daily living, it serves as an example of the need for more comprehensive views in global environmental management as the changes such as this can be detrimental.

***

Based on the clear findings, the Water Footprint Index (WFI) for cotton is rated "High," with the score of 3, while polyester is rated "Low," with the score of 1 in the Farming/Fiber-Making segment. If a precise numerical measurement method were applied to this criterion instead of the simple three step scoring system adopted for this LCA analysis with the maximum possible gap of 2, the difference between the two fibers would likely be in the range of tens of times, clearly illustrating the significant disparity in water usage.

11.3 Food Footprint Index (FFI) Analysis - Farming / Fiber-Making

All natural fibers require large areas of agricultural land for growth and harvest. In contrast, synthetic fiber production does not necessitate such land use. While some large manufacturing facilities for synthetic fibers might be built on previously agricultural lands, potentially impacting food sources, this issue is more economic than intrinsic to the manufacturing process. In other words, they can be constructed without sacrificing agricultural land, leading to fundamentally different impacts on agriculture compared to the cultivation of natural fibers.

Agricultural Land Use

Cotton, for instance, thrives in climates conducive to prolific agricultural outcomes, where multiple rounds of harvest for other crops can be achieved in a single season. In addition, as discussed earlier, cotton is a particularly water-intensive crop, demanding significantly more water than other food crops like soy, rice, wheat, potatoes, etc.

Replacement of Synthetic Fibers with Cotton: If we were to replace the current 65% of the total fiber consumption of the synthetic fibers with the natural fibers like cotton, as proposed by many environmental interest groups and activists, assuming a one-to-one substitution ratio, would necessitate tripling the agricultural land dedicated to cotton. Consider this: the world's land area is 13,003 million hectares, with 4,889 million hectares classified as agricultural by the Food and Agriculture Organization (FAO) of UN, representing 37.6% of the total land area. The 2021/2022 agricultural season saw the production of 25.2 million metric tons of cotton, occupying about 2.5% of total agricultural land. To replace the synthetic fibers, this proportion would need to increase to 7.5%, assuming a direct land-use swap.

However, such a one-to-one land swap is impractical. As mentioned, cotton grows in some of the most agriculturally productive areas, which yield more outputs per unit area than less fertile regions. Moreover, the assumption of cotton replacing synthetic fibers on a one-to-one physical property basis is flawed. For instance, cotton is five to ten times weaker than polyester. To match the strength of polyester, five to ten times larger quantities of cotton would be required to compensate the considerably weaker strength, further increasing the agricultural land required.

Factoring in these variables, the cotton cultivation could potentially consume well over 10% of the total agricultural land if the synthetic fibers were to be replaced by cotton. Additionally, cotton's particularly high water and pesticide demands would exacerbate the environmental and agricultural impacts.

World in Hunger: Reports from the World Health Organization (WHO) and the Food and Agriculture Organization (FAO) indicate that, as of 2023, 10% of the global population lacks sufficient food. This underscores the critical importance of judicious land use and resource management in balancing fiber production and food security.

Fig.43 World Hunger Map 2023

828 million people live in hunger. One in 10 people around the world go to bed hungry each night. Of those affected by hunger, two-thirds are women and.....

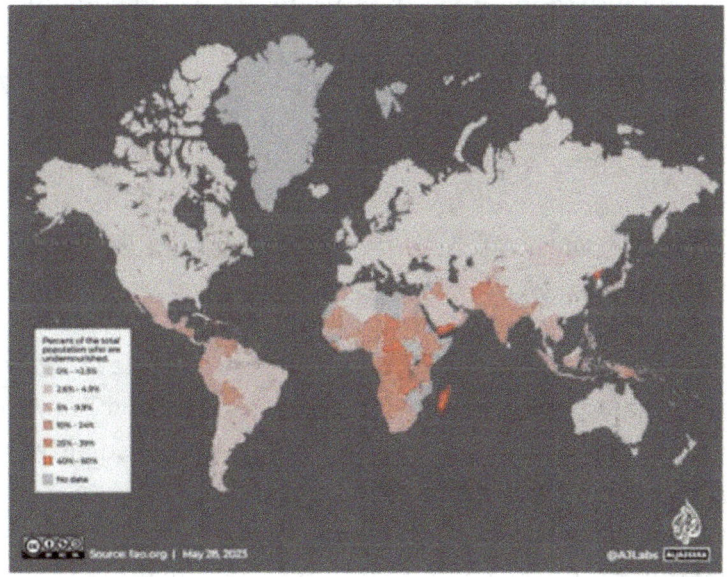

https://www.aljazeera.com/news/2023/5/28/why-is-global-hunger-on-the-rise-2

Considering the argument put forth by many environmental groups and activists that "*synthetic fibers, being derived from petroleum, are harmful to the environment and should be replaced with natural fibers"*, we must ask: Would the 828 million malnourished individuals worldwide agree with this substitution, knowing that it could further compromise their already limited food sources? In reality, if we were to follow these recommendations, a significant number of those advocating for such changes might find themselves facing food shortages as well.

Human Nature and Near-Certain Outcome: Throughout human history, it has been proven time and time again that, while the affluent often have an abundance of food, frequently discarding large quantities, the impoverished struggle to meet their basic nutritional needs. As global food security becomes more problematic, this disparity in food distribution becomes even more pronounced. If we were to replace agricultural lands used for growing food crops with cotton fields to swap polyester with cotton, I estimate that more than half of the world's population would struggle with food scarcity. This shift could most certainly lead to conflicts and exacerbate the global instability.

Perplexities of Wrongful Environmental Actions: Many of the largest non-governmental organizations (NGOs) today are environmental groups, primarily based in advanced economies such as Europe and North America. These entities, along with influential voices from politicians, governments, crown corporations, and private companies in developed countries, often advocate for sustainability measures without considering even the most basic scopes such as the food security challenges faced by many underdeveloped and developing nations.

Unfortunately, this disparity underscores a significant disconnect in the global sustainability management. It is truly perplexing that such claims and actions are made without addressing the basic needs of others. Effective and equitable sustainability practices must consider the diverse and urgent needs of all communities, ensuring that environmental goals do not exacerbate the existing inequalities and hardships.

Moreover, when we examine the environmental impacts and sustainability of the natural fibers in comparison with the synthetic fibers, it becomes evident that there is no justification for such illogical actions.

***

With the preceding considerations, cotton is rated as "High" with the score of 3 in Food Footprint Index (FFI) whereas Polyester receives the "Low" and 1 score.

11.4 Carbon Footprint Index (CEI) Analysis - Farming / Fiber-Making

One of the primary arguments made by many environmental interest groups and activists against synthetic fibers is the petroleum-based source materials and their inherent carbon footprints. While the cotton cultivation relies on natural sunlight, the polyester production requires energy from various sources. Although a small portion of this energy may come from the renewable sources, such as solar or wind, the majority still derives from the traditional resources like petroleum, coal, water, and natural gas. This further strengthens the claims of the environmental groups and activists.

However, a more nuanced analysis reveals that the issue is not as straightforward as merely considering the source materials and the energy need of polymer production. For instance, cotton cultivation involves the production and application of large quantities of chemicals such as pesticides and fertilizers. In addition, operations of manpower, machinery, etc., contribute significantly to its overall carbon footprint. Thus, the process of growing cotton is far from being carbon-free.

Scientific Analysis of Carbon Footprint Comparison: Understanding the overall carbon impacts of different fibers involves numerous factors. To provide an objective comparison, I refer to a research paper that employs comprehensive methodologies, as illustrated in *Fig.44* and the provided website link.

According to this report, the production of 1 pound of cotton fiber generates approximately 1.2 kg (2.7 lbs) of CO_2, while the production of polyester fiber results in about 1.1 kg (2.4 lbs) of CO_2. This close similarity may surprise many readers and challenges the common perception that synthetic fibers are inherently more carbon-intensive than natural fibers. In fact, the research highlights that some natural fibers, such as wool and silk, have their carbon footprints that are five to more than twenty times higher than that of polyester.

Fig.44 Carbon Footprint Comparisons

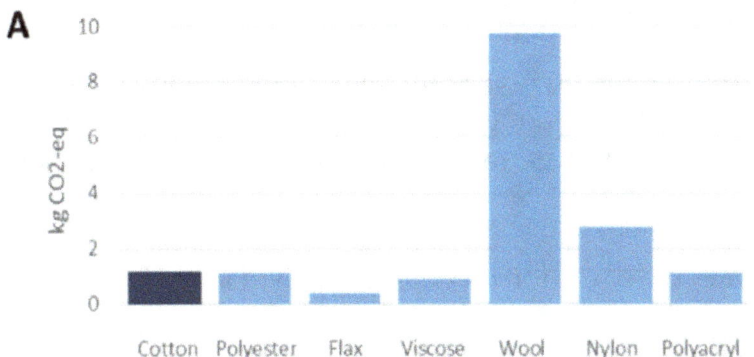

https://www.researchgate.net/figure/Impact-of-the-production-of-264-g-cotton-polyester-flax-viscose-wool-Nylon-and_fig2_349651454

* Special mention - It is crucial to approach this type of information with caution. This advice applies not only to this specific data set but to all scientific data, regardless of the source. Given the multitude and complexity of the factors involved, such as weather conditions, farming methods, and regional differences, it is virtually impossible to calculate a carbon footprint that accurately represents the global cotton population. Consequently, there are inherent limitations in this type of scientific data, and it is essential to understand the specific context and nuances of each data set.

***

Despite these limitations and considering the comparable carbon footprints between cotton and polyester as well as relatively lower values compared with other fibers reported in the research paper, I rate both fibers as "Low," with the score of 1 on the Carbon Footprint Index (CFI) for the Farming and Fiber-Making segment.

11.5 Total Impact Index (TII) Analysis - Farming / Fiber-Making

Based on the analyses conducted throughout this section, comparing the farming of cotton and the fiber synthesis of polyester, *Fig.45* represents the overall environmental impacts.

Fig.45 LCA – Cotton vs. Polyester – Farming / Fiber making

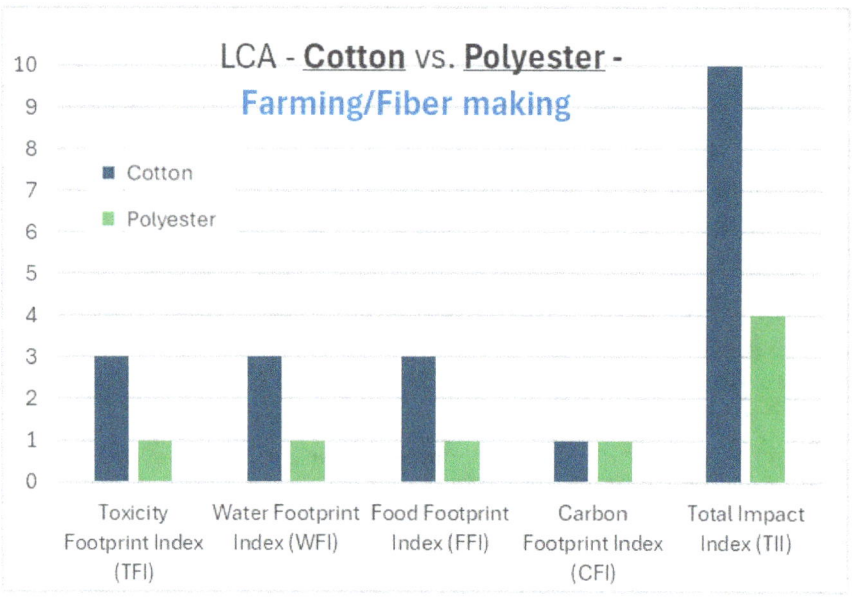

While polyester earned the "Low" score of 1 in all evaluation criteria at this initial stage of its lifecycle, cotton received the "High" score of 3 in three of the four areas, including the Toxicity, Water, and Food Footprint Indexes. The final score for cotton, 10 out of the maximum possible score of 12, starkly contrasts with the lowest possible score of 4 for polyester.

In summary, the results unequivocally demonstrate that cotton exhibits significantly higher levels of negative environmental and also humanitarian impacts from the food security perspective in the Farming and Fiber-Making segment.

Section 12 LCA – Manufacturing

12.1 Toxicity Footprint Index (TFI) Analysis - Manufacturing

All manufacturing processes for textile goods leave an environmental footprint, just like any other human activity. For example, operating a sewing machine requires energy, and although the energy consumption of a small sewing machine is minimal, it still contributes to the overall environmental impact. In this section, my focus is not on heavily critiquing manufacturing processes with relatively minor environmental footprints. Instead, I concentrate on the most significant causes and impacts that are more detrimental to the environment.

Chemical Toxicity in Dyeing Process: As previously emphasized, dyeing is the most toxic process in the manufacturing of most textile goods, excluding the farming of natural fibers. The toxic pollutants generated during the dyeing spread globally through air and water circulation, creating long-lasting impacts on the health of all living organisms.

Earlier in the book, specifically in **Chapter II, Section 10.6 Dyeing**, I discussed the variety of chemicals used in the dyeing processes, including dye pigments, whitening agents, fixing agents, mordants, etc. Many of these chemicals contain heavy metals, aromatic amines, and other harmful compounds, either as primary ingredients or as the byproducts of various chemical reactions. These substances are crucial in many dye formulations for achieving vibrant colors consumers expect and for ensuring that these colors remain stable on the textiles during use. In other words, the colors would neither meet consumer preferences nor retain their vibrancy over time without these harmful chemicals.

Toxicity Comparison between Natural and Synthetic Fibers: During my hands-on dyeing experiences, one notable observation I made was the significant disparity in the chemical usage between dyeing natural and synthetic fibers. Achieving a certain color tone and hue in natural fibers similar to that of synthetic fibers often required multiple times higher quantities of chemical usages. Moreover, the dyes and associated chemicals employed for natural fibers are inherently more chemically reactive, where higher chemical reactivity leads to more harmful environmental impact. In contrast, the disperse dyes used in most synthetic fibers are chemically inert.

Other Areas of Manufacturing: Outside the dyeing process, various levels of toxicity can be introduced in different manufacturing stages to achieve specific performance features. For example, highly toxic chemicals are used to impart flame-retardant properties to cotton fabrics, particularly for specific types of personal protective equipment (PPE), where the property is essential to perform highly dangerous job functions such as fire-fighter gear. Given the focus of this book on high overall impact areas, I will not delve into the niche areas of manufacturing as they constitute a very small portion of the overall textile manufacturing while the needs for these specialized chemical treatments are often unavoidable.

***

In consideration of these factors, cotton is assessed with "High" in Toxicity Footprint Index (TFI) in the Manufacturing segment, gaining the score of 3. In contrast, polyester garners "Low" rating with the score of 1, reflecting its comparatively lower environmental impact in terms of the chemical toxicity during its manufacturing processes.

12.2 Water Footprint Index (WFI) Analysis - Manufacturing

For textile manufacturers, water usage relates closely with the overall operational efficiency and cost management. The more water used, the more energy required to dry, thus more costly. In fact, the energy required for drying water-heavy materials is often one of the largest operational costs for dyeing facilities.

Increased quantities of the chemical use and higher chemical reactivity, which are closely interrelated, for the natural fibers like cotton significantly contribute to higher water consumption in comparison with the synthetic fibers like polyester. Thorough rinsing, as previously explained with the analogy of hair dyeing, is crucial, especially given the harmful nature of these chemicals. Consequently, natural fibers require substantially more water during the dyeing process to effectively remove excess chemicals.

> * Special Mention – Wool: Animal-based fibers, such as wool, present distinct challenges in their manufacturing processes, particularly due to their extensive cleaning requirements. For instance, sheep, the primary source of wool, often come into contact with their own manure, resulting in contaminated

fibers at time of the harvest. This necessitates thorough cleaning before using them in subsequent processes.

During my tenure in South Korea, I recall that the company sold some byproducts from their wool cleaning process as the wastewater contained oil residue extracted during washing, valuable for some cosmetic products, highlighting an eco-friendly aspect of upcycling within this cleaning process.

Despite these beneficial side-effects, the initial cleaning of wool demands a significant amount of water and chemicals to effectively wash off the contaminants. Moreover, compared to plant-based fibers, animal-based fibers necessitate even higher quantities of dyes and chemicals. This increased demand is attributed to the protein-based molecular structure of animal fibers, resulting in higher densities of chemically reactive functional groups in their molecular structures. These functional groups not only enhance the affinity for water but also require more extensive use of harmful chemicals during the dyeing process. The manufacturing processes of the animal-based fibers, therefore, entail even higher water consumption and heightened use of toxic chemicals than the plant-based fibers, underscoring the environmental challenges associated with these materials in textile production.

***

Given these considerations, cotton is assigned as "High" impact with the score of 3 in the Water Footprint Index (WFI) for the Manufacturing segment. In contrast, Polyester receives "Low" rating with the score of 1, indicative of its low water footprint driven by its inherent low chemical reactivity.

12.3 Food Footprint Index (FFI) Analysis - Manufacturing

Impacts on the food footprint in textiles are created mostly during the farming segment of the natural fibers. Therefore, the manufacturing of textile goods using either cotton or polyester does not create any direct impact on the food securities. With this, both cotton and polyester receive the score of 1 for "Low" in the Food Footprint Index (FFI) during the Manufacturing segment.

12.4 Carbon Footprint Index (CFI) Analysis - Manufacturing

The disparity in the carbon footprint between natural and synthetic materials is pronounced particularly in the Manufacturing segment, primarily attributable to the dyeing process. Natural fibers exhibit higher water absorption during dyeing, which subsequently escalates the energy consumption in drying.

While a direct experience in a dyeing factory may be limited for most readers, a comparative estimation can be made at home - Conceptually, the carbon footprint impacts of the dyeing operations in factories can be loosely interpreted as higher magnitudes than those in the home laundry as both involve the usage of large quantity of water and rinsing operations. To illustrate this, I present a set of data derived from a controlled experiment conducted in a domestic setting, outlined in **13.4 Carbon**

Footprint Index (CFI) – Consumption section later in this chapter.

***

Given that the Carbon Footprint Index (CFI) is closely related with the Water Footprint Index (WFI), cotton is evidently scored with 3 as "High" and polyester as "Low" acquiring the score of 1 for the Manufacturing segment of the LCA.

12.5 Total Impact Index (TII) Analysis - Manufacturing

Summing up the scores of the four separate indexes in the Manufacturing segment, cotton has the final score of 10 out of the highest possible score of 12. Cotton received the highest scores in all the other segments but the Food Footprint Index (FFI), as the criteria excludes the Farming, whereas polyester receives the total score of only 4, registering the lowest score for all indexes in the same segment.

Fig.46 LCA – Cotton vs. Polyester - Manufacturing

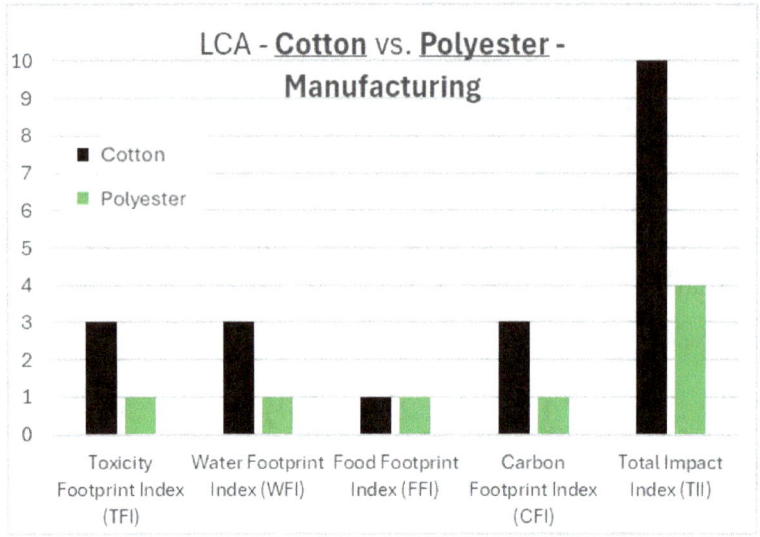

Section 13 Consumption

Having explored the environmental impacts associated with the earlier stages of a textile's lifecycle, namely, the farming for natural fibers and the fiber synthesis for synthetic materials, as well as the manufacturing processes, this section will now shift focus to the environmental considerations during the consumption phase of textile goods.

13.1 Toxicity Footprint Index (TFI) Analysis - Consumption

Before scoring the impacts associated with the toxicity of textile goods during consumption, already discussed throughout the book in various related subjects, I present short summaries below.

Chemical Absorption and Release: Natural fibers absorb and release higher quantities of chemicals due to their heightened chemical reactivities of the fibers themselves as well as the chemicals used in farming and manufacturing. Furthermore, *Fig.14* Breaking Energy Comparison (between cotton and polyester) and related discussions revealed clear scientific evidence in the higher quantity of dust generated from natural fibers in comparison to synthetic ones. Finally, natural fibers incur higher probability of drycleaning as extensively discussed in **1.1.1 Care Method Overview.**

Combining these metrics, the natural fibers present unequivocally more significant environmental challenges during consumption.

Persistence in Nature: While some environmental advocacy groups present the decomposition speed between natural and synthetic fibers as a main argument in their environmental sustainability evaluations, it is critical to note that some of the chemicals absorbed in these materials will last multiple times longer than the fiber materials in their pure forms, including synthetic fibers; more in-depth discussion will be presented in the **LCA – Disposal** section of this chapter.

> * Special Mention - Personal Shopping Habits and Maintenance Routines: I avoid purchasing items generally considered as "slow fashion" for several reasons.

Firstly, with my background, it is not difficult for me to assess the cost-quality ratio and recognize the significant markups associated with "slow fashion" brands. A stark difference between manufacturing costs and retail prices discourages me from paying a premium to support high executive salaries, luxurious office spaces, marketing expenses, costly retail locations and operations of these brands.

Secondly, more affordable textile items, often made from synthetic fibers, offer superior durability without the need for drycleaning. While these items might be more prone to minor issues like losing buttons or seam breakage, the overall strength and longevity of the synthetic fabrics are undeniably superior to those of natural fibers.

Thirdly, and most importantly, from an environmental perspective, I understand that synthetic materials offer significant advantages in most aspects of their lifecycles.

In my care routines, I sort laundry by dark and light colors, much like most people, and avoid overloading the washing machine to prevent excessive physical stress inside the machine. If buttons fall or seamlines break, I bring them to tailors. By following these routines, I maintain my textiles in good condition, allowing them to last for extended periods without the need for dry cleaning.

***

Given these considerations, it is evident that cotton generates significantly higher environmental harm during consumption, earning it "High" rating with the score of 3, while polyester is rated "Low" with the score of 1.

13.2 Water Footprint Index (WFI) Analysis - Consumption

As previously established, natural fibers, such as cotton, exhibit a higher affinity for chemical reactions compared to synthetic fibers like polyester. Conversely, cotton can absorb much more quantity of water than polyester during laundry. This disparity in water absorption is readily apparent when comparing the weights of cotton and polyester garments during laundry, more of which will be discussed in the subsequent subject with an experiment designed to illustrate the difference.

Moreover, cotton's higher affinity to water can contribute to easier odor development. High moisture contents of natural fibers provide habitats for living organisms and human sweat containing food source for them such as urea, ammonia, fatty acid, etc., facilitate bacterial growth. This, in turn, necessitates more frequent washing cycles for cotton than polyester.

***

Given these considerations, cotton receives the score of 3 for "High", while polyester receives 1 for "Low" in the Water Footprint Index (WFI) for the Consumption segment.

13.3 Food Footprint Index (FFI) Analysis - Consumption

Consumption of textile goods using either cotton or polyester does not create any direct impact on the food securities. Thus, both cotton and polyester receive the score of 1 for "Low" in the Food Footprint Index (FFI) of the Consumption segment.

13.4 Carbon Footprint Index (CEI) Analysis - Consumption

As previously established, cotton cultivation requires large amounts of water due to the inherent hydrophilicity of its cellulose structure. This water affinity persists throughout the lifecycle, necessitating more water usage during washing and rinsing cycles and increasing energy consumption in the drying process of home laundry.

A home-based experiment was conducted to compare the energy consumption of laundering between cotton and polyester. The experiment is aimed to mimic real-life laundry scenarios with the detailed methodology presented as follows.

Preparations:

- Two groups of towels as the test samples:

- Group 1: Four towels made with 100% polyester; each towel sized at 69 x 132cm (27 x 52 inches)
- Group 2: Four towels made with 100% cotton; each towel sized at 69 x 132cm (27 x 52 inches)

- Washer make and model: Blomberg WM77120 NBL01, Front loading
- Dryer make and model: Blomberg DV1542 model, 15 Amp, 2,000 Watt, 3.7 cu.ft capacity, Front loading
- 30kg x 1g Electronic Computing Scale

Test Procedure:

1. Each group of the four towels is placed in a room with temperature controlled at 23°C (or 73°F) and relative humidity between 50 ~ 60% for a minimum of 24 hours, a process generally known as "conditioning" in typical laboratory tests.

2. After the conditioning, each group of the four towels is measured in weight. This initial weight is identified as the "DRY (weight)" in **Fig.47**.

3. Each group is washed separately in the same setting of the washing machine, which includes the spinning at the end to remove excess water

4. Once the washing cycle is complete, the weight of the four towels in the same group is measured and marked as "WET (weight)" in the result.

5. During the drying cycle, the weight of the four towels in the same group is measured in 10 minutes interval until it reaches the initial DRY (weight).

The results of the experiment are as follow:

Fig.47 Towel Wash Experiment Results - Table

	Group 1 Weight (KG) POLYESTER 100%	Group 2 Weight (KG) COTTON 100%	Residual Water Weight (KG) POLYESTER 100%	Residual Water Weight (KG) COTTON 100%	Residual Water % POLYESTER 100%	Residual Water % COTTON 100%
DRY	1.416	1.875				
WET	1.915	3.075	**0.499**	**1.2**		
DRY 5'						
DRY 10'	1.61	2.767	0.305	0.308	61.12%	25.67%
DRY 15'	1.456		0.459		91.98%	
DRY 20'	1.395	2.421	0.520	0.654	104.21%	54.50%
DRY 25'						
DRY 30'		2.166		0.909		75.75%
DRY 35'		2.022		1.053		87.75%
DRY 40'		1.928		1.147		95.58%
DRY 45'		1.847		1.228		102.33%

* Notes: For readers' reference, it is natural to reach a lower weight than the initial DRY weight after drying because the initial weight includes moisture absorbed by textile goods during storage (or conditioning in the case of this test). The test extended to dry this pre-existing moisture. Fine adjustments to calculate the time required to reach the exact initial dry weight could have been made. However, the significant difference in the test results between the two groups did not justify such fine adjustments.

The Group 2 cotton towels retained 240% more water after the washing-spinning process, with the "WET" weight of 1.2 kg (or 2.65 lbs) compared to 0.499 kg (or 1.10 lbs) for the polyester towels. Additionally, the drying time for the cotton group was significantly longer, taking 45 minutes versus 20 minutes for the polyester group.

Converting the above figures to energy consumption:

- The Cotton Group consumed 1.5 kWh of energy (=2,000 watt/1,000 x 45/60)

- The Polyester Group consumed 0.67 kWh of energy (=2,000 watt/1,000 x 20/60)

This extended drying time for the cotton group resulted in an energy consumption difference of 0.87 kWh between the two groups (or 124% higher for the cotton). Considering that textile goods can be washed easily tens or even hundreds of times throughout their lifecycles, this difference translates to significantly different outcomes in the carbon footprint.

In a graphic format, the results are demonstrated in **Fig.48**.

Fig.48 Towel Wash Experiment Result - Graph

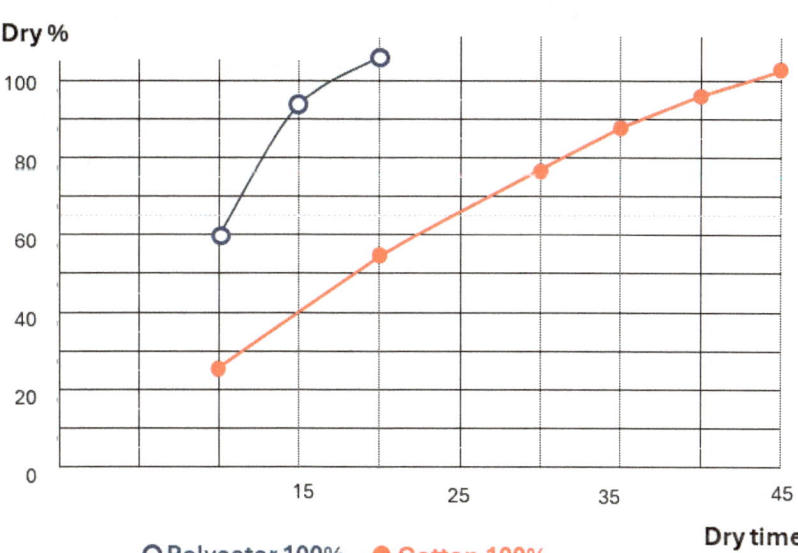

***

As the experiment results clearly indicate, cotton requires considerably more amount of energy during consumption. Based on these factors, cotton receives the score of 3 for "High", while polyester receives 1 for "Low" in the Carbon Footprint Index (CFI) for the Consumption segment.

13.5 Total Impact Index (TII) Analysis - Consumption

Summing up the scores, cotton gains the final score of 10 out of the highest possible score of 12. Cotton is rated with the highest scores in all indexes except for the Food Footprint Index (FFI) during the Consumption segment. In contrast, polyester receives the total score of only 4 while registering the lowest scores for all criteria in the segment.

Following the two previous segments, the Consumption revealed identical results, highlighting significantly higher level of negative environmental impacts from cotton in comparison with polyester.

Fig.49 LCA – Cotton vs. Polyester - Consumption

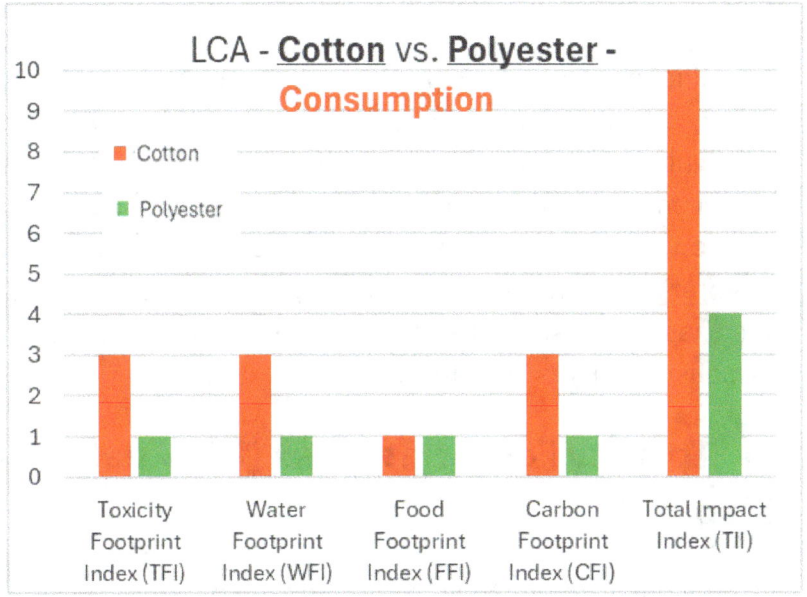

Section 14 Disposal

A significant disparity exists between the public perception and the reality of textile waste disposal. This discrepancy highlights broader challenges in environmental management and human behavior. While **Chapter I, Section 3: Unfortunate Disposal**, offered an in-depth exploration of improper disposal practices, this section analyzes these issues based on the principles and metrics established in this LCA analysis.

14.1 Toxicity Footprint Index (TFI) Analysis - Disposal

As previously established, the prolonged decomposition of synthetic fibers and microplastics is not the sole factor in assessing the environmental impacts of natural vs. synthetic fibers. While the cellulose in cotton biodegrades significantly faster than polyester, the residual chemicals embedded in textile goods often persist in the environment far longer than any synthetic fibers.

Earlier discussions highlighted various factors influencing this analysis, including chemical compositions, reactivities, toxicities, persistence, and quantities of chemicals absorbed and released. These factors reveal significant differences between natural and synthetic fibers.

Despite scientific evidence suggesting that natural fibers generally have a greater environmental impact, there is another critical metric to consider. Although precise data is lacking, it is estimated that approximately 65% of the global textile waste is composed of various synthetic materials, with the remaining 35% being natural fibers - This rough estimate is based on the total fiber consumption presented in *Fig.1* and *2*, drawing parallels with the composition of the disposed materials.

Furthermore, according to the U.S. Environmental Protection Agency (EPA), the most recent report indicates that over 17 million tons of textile Municipal Solid Waste (MSW) are generated annually, which equates to around 50 kg (112 lb) per person for the entire U.S. population. While I always approach statistics with caution, this is an alarmingly high quantity.

Given the sheer quantity perspective and proportionally high residual chemicals in the waste, both cotton and polyester are rated as having "High" impact, each gaining the score of 3.

14.2 Water Footprint Index (WFI) Analysis - Disposal

There is mounting evidence of water pollution caused by textile waste, particularly in underdeveloped and developing countries. As previously discussed, large quantities of textile waste from affluent nations are shipped to these regions, where they are frequently abandoned in natural environments or disposed of without any environmental safeguards. Massive piles of waste are often found along riverbanks, seemingly left to deteriorate over time. The water in these rivers appears heavily polluted and lifeless. Additionally, the practices of burying and burning textile waste without proper environmental controls further contaminate runoff and groundwater.

While there are numerous serious concerns regarding water pollution in this disposal segment, making a precise comparison between natural and synthetic fibers is incredibly challenging due to the scale and complexity of such an analysis. However, it is clear that both types of fibers contribute significantly to environmental damage. Given this, I have assigned both cotton and polyester the "High" impact rating, with each receiving the score of 3.

14.3 Food Footprint Index (FFI) Analysis - Disposal

Textile waste can have numerous indirect impacts on various food sources. For instance, highly contaminated soil and water may either prohibit farming activities or result in the transfer of pollutants to other living organisms, including humans, through food sources. As with the analysis of the **Water Footprint Index (WFI) – Disposal** earlier, conducting a comprehensive analysis yielding accurate results in the food impacts between natural and synthetic is nearly impossible.

However, underestimating the negative impacts of textile waste in our food sources would be a significant mistake, given the chemical quantity, toxicity, and reactivity remaining in the waste as well as the sheer volume of the waste. While all indications clearly suggest that natural fibers have a more negative impact than synthetic fibers, the degree of the unknown factors prevents me from assigning different scores to the two. Consequently, both cotton and polyester are rated as having "High" impact, each receiving the score of 3.

14.4 Carbon Footprint Index (CFI) Analysis

The decomposition behaviors of natural and synthetic fibers are influenced by different mechanisms driven by microbial activities. In natural fibers, bacteria and fungi break down organic molecules like cellulose and lignin in plant-based fibers, and keratin and other proteins in animal-based fibers, into simpler compounds. This decomposition process generates significant amounts of carbon emissions. Specifically, aerobic decomposition in the presence of oxygen produces CO_2, while anaerobic decomposition in environments like landfills releases CH_4 through microbial metabolism. Due to their chemical compositions and faster decomposition rates, natural fibers tend to generate more carbon emissions compared to synthetic ones. For instance, a study shows that cotton releases 15% more CO_2 compared to polyester when decomposed aerobically.

Simultaneously, the decomposition of plastics in the environment is an area of ongoing research. While there is clear evidence of plastic-eating bacteria and the natural breakdown of plastics, this field remains relatively underexplored. Some studies have assessed the carbon impact of specific plastics, like Polylactic Acid (PLA), but these studies often have limited scopes and do not comprehensively address other plastic types.

To ensure that no critical factors are underestimated in this analysis, the uncertainty surrounding the plastic decomposition process does not prevent me from assigning both cotton and polyester the "High" impact score of 3 each.

14.5 Total Impact Index (TII) Analysis - Disposal

My decision for scoring "High" in all the indexes in the Disposal segment for both cotton and polyester reflects the following consideration - It is estimated that trillions of textile goods are currently in use, all destined for eventual disposal. Additionally, tens of billions of textile items are produced annually and will continue to be produced in extremely large quantities. The residual chemicals contained in the current and future textile waste create detrimental environmental impacts in terms of toxicity, food, water, and carbon footprints.

With these considerations, attempting to evaluate the impact of each fiber type in the Disposal segment is meaningless in my view, explaining the simplistic approach of assigning the "High" scores for all the evaluation criteria.

Fig.50 LCA – Cotton vs. Polyester – Disposal

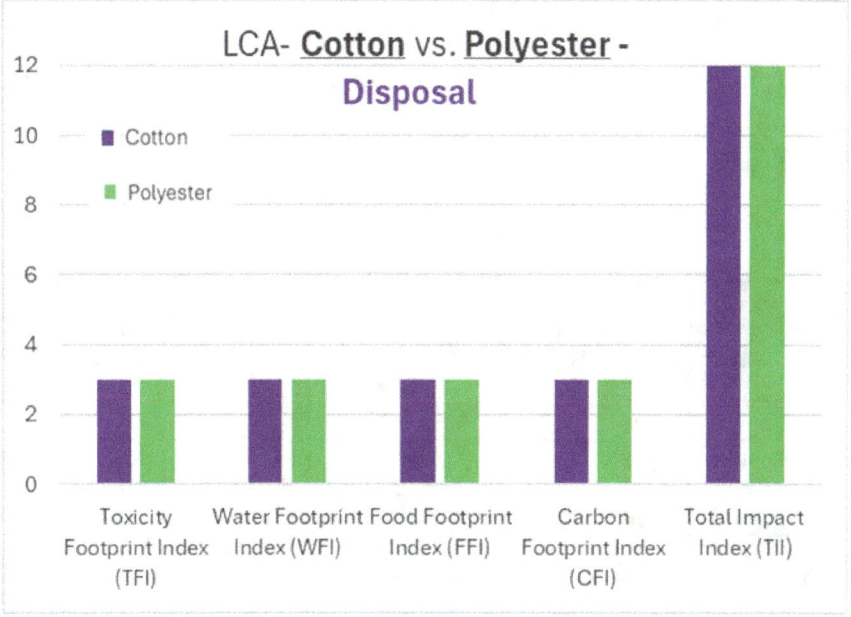

Section 15 LCA – Total Impacts

The heatmap and the final scores of the LCA results between the cotton and the polyester are presented in ***Fig.51*** and ***Fig.52*** as the LCA – Total Impacts.

Fig.51 Overview of LCA – Cotton vs. Polyester

Evaluation	Farming/Fiber making		Manufacturing		Consumption		Disposal		Highest score	Total	
	Cotton	Polyester	Cotton	Polyester	Cotton	Polyester	Cotton	Polyester		Cotton	Polyester
Toxicity Footprint Index (TFI)	3	1	3	1	3	1	3	3	12	12	6
Water Footprint Index (WFI)	3	1	3	1	3	1	3	3	12	12	6
Food Footprint Index (FFI)	3	1	1	1	1	1	3	3	12	8	6
Carbon Footprint Index (CFI)	1	1	3	1	3	1	3	3	12	10	6
Total Impact Index (TII)	10	4	10	4	10	4	12	12	48	42	24

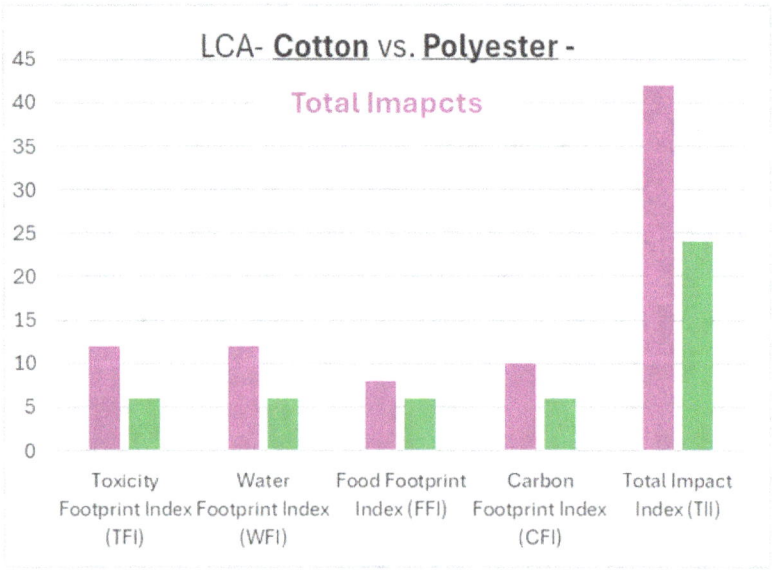

Fig.52 LCA – Cotton vs. Polyester – Total Impacts

Limitations: The index system employed in this analysis provides a valuable framework for the comparative evaluation of environmental impacts between the cotton and the polyester fibers as the representatives of the natural fibers and their synthetic counterparts respectively. As mentioned earlier, however, it is acknowledged that a more granular quantification of the real-world impacts would likely amplify the disparity between the two fiber types across various criteria.

For instance, a direct volumetric comparison of water usage for the Water Footprint Index (WFI) calculation could potentially yield a much more pronounced difference, possibly by a factor of 10 or even more, as the polyester does not require the extensive water usage in farming while consuming considerably lower quantities during the manufacturing and consumption. However, this analysis shows a numerical difference of only 2 because the predetermined method employed a simplistic approach of High, Medium, and Low, assigned with the scores of 3, 2, and 1 respectively – Previously established was the inherent complexities and variabilities throughout the lifecycles of different fibers rendering more precise numerical quantification extremely challenging, if not infeasible.

Substantial Differences: Notwithstanding these limitations, the substantial difference in the Total Impact Index (TII) scores of 42 for the cotton versus 24 for the polyester is indicative of significant environmental implications between the two fibers.

While all 16 evaluation areas are critical for our sustainable future, the on-going humanitarian crisis with the shortage of food in nearly 10% of the world population prompts an immediate attention in relation with many environmental advocacy groups and activists favoring the natural fibers over the synthetics. It is clearly our humanitarian duty to alleviate such crisis and any argument on the environmental sustainability without addressing such issue needs to be scrutinized.

Chapter III Post-Chapter Commentary

While the Life Cycle Assessment (LCA) results presented in this chapter highlight clear advantages of polyester over cotton, it is essential to acknowledge the complex and nuanced nature of material selection decisions made by the public. Each material choice inevitably comes with both benefits and drawbacks. Moreover, certain factors, such as tactile preferences, consumers' long-standing emotional attachments, and the socioeconomic impacts on the natural fiber farmers worldwide, were not fully captured in this LCA analysis.

Many of these factors represent important dimensions of our lives, that cannot be ignored nor easily altered. While I have had a firm belief in the environmental benefits of synthetic fibers over natural ones and have reflected it in my purchasing decisions wherever applicable, I do not expect such conviction will transfer to others, significantly changing their shopping patterns in favor of synthetic materials, nor should they, as there may be too many complex socioeconomic implications.

Rather, the emphasis I intend to address is the unscientific nature of such simplistic views as *"natural fibers are more sustainable than synthetic fibers"* without a valid scientific ground. In the intricate fabric of our modern lives, the choices we can make within given circumstances are often limited, and material choices are some of the examples. However, more scientific approaches can lead humanity to make more favorable decisions in the mid- to long-term and help us transition to better solutions for the environmental sustainability in a reasonable timeframe.

Chapter IV

****

Misinformation Crisis

Pre-Chapter Commentary

Quote of the chapter

"A truly incomprehensible flaw of this program* is its inability to prevent the inclusion of microorganisms in the used goods deposited by random consumers. Various microorganisms, such as bed bugs, dust mites, bacteria, etc., can be present in used textile goods." - * Greenwashing Case #2 presented in this chapter

***

In the previous chapters, I shared many potentially controversial views which may contradict the general public's existing perception on the environmental impacts of certain textile materials. In doing so, I presented relevant scientific data and practical experiences I have gained both professionally and through personal observations.

The significant gap I recognize between the common notions and the information presented in this book arises from widespread misinformation on the internet. The easy access to unfiltered and scientifically unsupported information has profoundly influenced global citizens' perspectives on the environmental sustainability.

In this chapter, I will explore specific examples of such phenomenon and examine how the public is often misled, either by the lack of knowledge and expertise of those spreading the information or through purpose-driven actions by various stakeholders seeking commercial, political, or operational benefits.

Section 16 Greenwashing

Definitions – Cambridge Dictionary

a. to make people believe that your company is doing more to protect the environment that it really is.

b. an attempt to make people believe that your company is doing more to protect the environment than it really is.

c. an attempt to make your business seem interested in protecting the natural environment, when it is not.

While the definitions provided by the Cambridge Dictionary offer a conceptual understanding of "greenwashing", I believe they lack several critical executable aspects relevant to real-life situations. This gap may create numerous loopholes in the way it is used and how many organizations exploit the public's genuine desire for cleaner environment. In this section, I will expand on these definitions by incorporating the following additions:

d. Failing to provide full disclosure behind specific sustainability claims

e. Failing to include comprehensive aspects of claims, including potentially negative outcomes

f. Relying on external certification programs driven by a lack of internal expertise in the related areas.

g. Misleading the public with misinformation

Proper Conduct for Sustainability and ESG Efforts

While several real-life greenwashing examples that may violate the definitions will be discussed in separate case studies later in this section, it would be beneficial for readers to review certain criteria built upon the original and extended definitions in the following 5 areas to gain more practical understanding on honest and earnest sustainability and ESG (Environmental, Social, and Governance) effort and to distinguish it from potential greenwashing.

1. **Full Disclosure**: Interest groups or individuals must provide complete and transparent information regarding their sustainability claims and actions. This includes offering scientifically valid data coherent with the intended use of the information, any known and potential side effects associated with it.

2. **Comprehensive Scope**: Parties of interest must present potentially negative outcomes beyond the scope of their claims and actions. This includes making earnest efforts to understanding and sharing the global impacts of their actions.

3. **Internal Expertise**: Claim-makers are responsible for developing adequate internal expertise related to their claims. Solely relying on external certification programs or 3rd party authorities does not absolve them from the responsibility of substantiating their claims accurately with valid science.

4. **Accountability**:

    Knowingly Misleading: When a claim is made or an action is initiated with elements of intentional greenwashing, it constitutes "organizational dishonesty".

Unknowingly Misleading: When a claim is made or an action is initiated without sufficient knowledge, it reflects "organizational incompetence".

Both scenarios represent corporate misconduct and greenwashing.

5. **Avoidance of Misinformation**: Organizations and individuals must refrain from promoting information that lacks scientifically valid evidence. This is especially critical in today's context, where the public interest in environmental issues is exceptionally high. Unfortunately, many environmental groups, including government bodies, NGOs, and activists, are involved in spreading such misleading information.

By integrating these principles into the metrics used to evaluate various sustainability claims and actions, we aim to create a more robust framework for assessing the integrity of their actions. Clearer definitions and expectations will help prevent greenwashing and encourage genuine environmental stewardship.

Case #1 - A global sportswear company states that a certain percentage of their merchandise is made from recycled content, with the recycled fibers certified by a third-party certification program.

These claims are often accompanied by additional information, such as how many plastic bottles were recycled to produce a specific garment like a shirt or jacket.

To determine whether such claims constitute greenwashing, two primary aspects must be scrutinized: the nature of the recycling processes used, and the transparency and validity of the certification programs cited. Many readers might be surprised by the extent of the loopholes that can exist in "recycled" claims. In fact, I have yet to encounter a claim of this kind that fully meets the disclosure requirements necessary to validate its legitimacy as truly beneficial to environment. Given the potentially surprising nature of this statement, I will share the journey I undertook to investigate the reality behind conventional plastic recycling practices.

Pursuit of Sustainable Recycling

My interest in plastic recycling methods emerged while seeking sustainable product options for one of the businesses I manage. Adopting a third-party certification program, as many companies do, would have been straightforward and would not have delayed product launches. It merely required instructing my manufacturing partners to purchase recycled fibers with a widely recognized certificate readily available in the market. However, I refrained from this approach for three reasons.

Contamination: Firstly, I observed that a large portion of the plastics collected for recycling were quite dirty, with visible contaminants such as dirt, soil and liquids. I noticed this in community recycling bins where contaminated plastics were deposited in most countries I traveled to. With my engineering background, I naturally became curious about how these pollutants were removed during recycling processes.

Ban on Plastic Waste: Secondly, China's 2018 ban on importing plastic waste, followed by many other underdeveloped and developing countries around the globe, piqued my interest. While consumers in the Western world often view recycling as virtuous, I was curious about why these countries chose to ban these supposedly virtuous material sources.

Personal Commitment: Thirdly, I wanted to ensure that my products were genuinely eco-sustainable by personally verifying the recycling processes rather than relying solely on external certifications. This determination stemmed from many occasions of unrealistic and even contradictory certification programs witnessed in my professional life in the industry.

To gain firsthand insight into the processes, I visited many recycling and manufacturing facilities certified by globally recognized certification programs. Unfortunately, my investigation led me to the conclusion that conventional recycling methods often result in significant negative environmental impacts, which can easily outweigh the benefits. The detailed scope and findings of this investigative work were provided in **Chapter I: Section 4 Unfortunate Recycling**.

Consequently, I decided against launching a recycled product based on the conventional recycling method as a sustainable product option in my business.

Poor Business Practices

It is troubling to see many companies, including some of the largest and most reputable global brands, claiming to use recycled fibers without addressing the overall environmental impacts of recycling processes or merely citing third-party certification programs.

Easily Detectable Problems: Anyone with a basic understanding of sustainable manufacturing practices visiting these facilities would immediately notice several areas of major concern: 1. Initial contamination of plastic waste as the source material, 2. Generation and disposal of microplastics into water from the crushing and grinding process mechanisms, 3. Significant amount of water and energy required for cleaning, 4. Residual contamination and its impact on the strength of recycled fibers and 5. Inefficient use of harmful chemicals associated with subsequent dyeing process.

Lack of Efforts: Encountering the claims of recycled contents without addressing such concerns leads me to believe that the organizations in question made no effort to understand the nature of their claims. Unfortunately, this aligns with the typical definition of greenwashing. My observations often reveal that even some of the largest brands, which have ample resources and operational capabilities, avoid fulfilling these responsibilities. This practice undermines genuine environmental efforts and misleads consumers regarding the true sustainability, or lack thereof, of their products.

Loopholes: A related topic in this discussion is the significant loopholes present in third-party certification programs. Those unfamiliar with the intricacies of these programs might assume that comprehensive efforts were made by the involved parties and that the associated claims fully account for overall environmental benefits, including potentially negative impacts. However, this is often not the case. Given the importance of this issue, a detailed discussion on the loopholes in the certification programs will be presented later in this chapter, in **Section 19 Loopholes in Certification Programs**.

***

Given the scope, I judge any "recycled" claim without clearly addressing these concerns as greenwashing.

Case #2 - A large fashion company launches a sustainability program, encouraging consumers to bring used garments to their retail stores and place them in the recycling bins in their stores. The company claims these garments are cut and upcycled as thermal insulation.

People unfamiliar with the engineering principles behind thermal insulation materials might assume it is a great idea and a sustainable choice. However, this perception is fundamentally flawed. As the concept of this claim relates to a traditional insulation material and technology for apparel, the following explanations are in the scope of such.

Lack of Understanding: Effective thermal insulation requires a solid structure to consistently and efficiently capture the body heat. Creating such a structure begins with well-engineered fibers that have desirable attributes, such as optimal lengths, thicknesses, etc., and are processed under manufacturing conditions that allow them to intertwine with each other and create adequate quantities and desirable sizes of air pockets.

However, the mechanical processes of chopping and slicing the deposited textile goods produce a significant amount of dust and very short fibers. When fibers are too short, they tend to float rather than form an integral structure to create the air pockets. During normal wear and laundering, this floating dust and short fibers escape from the enclosing structures. As a result, these materials provide little to no insulating value and are not effective as thermal insulation.

Microorganismal Flaw: A truly incomprehensible flaw of this program is its inability to prevent the inclusion of microorganisms in the used goods deposited by random consumers. Various microorganisms, such as bed bugs, dust mites, bacteria, etc., can be present in used textile goods. Re-emergence of some pests in large urban centers, once believed to be extinct from developed regions, is related to the rapid population growth and their centralization as well as the banning of toxic chemicals like DDT as pesticides in many countries many years ago.

Detecting microorganisms in used textiles would require examining every square inch of the used goods under a microscope. This process would incur an extremely high cost and labor. Even if the company making this claim implemented a sophisticated machine vision system for this purpose, it would not be a realistic solution because the microorganisms would spread rapidly to other garments in the deposit boxes and during processing. Circumstantially, the only viable remaining option would be to use toxic pesticides to eliminate these organisms.

Two Bad Options: Once applied, residual pesticides cannot be easily washed off due to their chemical durability, and washing would also result in losing a large portion of short fibers and dust. Therefore, the overall outcome of this program leaves consumers with one of two options: a collection of dust containing microorganisms or a collection of dust containing pesticides.

***

As an engineer with several breakthrough technologies and multiple patents in my textile career, I do not dismiss the possibility that the company associated with this claim has developed an innovative processing method to overcome these challenges. However, if such a method existed, one would expect the company to promote it to highlight their achievements. This promotion would enhance their corporate image as a genuinely green and sustainable company and inspire other companies in similar sustainability efforts. Unfortunately, no such information was available at the time of writing.

In my view, this level of greenwashing is truly astonishing as it lacks even the most fundamental understanding of the related matters of the claim, yet such practice is unfortunately not uncommon in the global textile industry and consumer markets.

Case #3 - Another global sportswear company announces an "enzyme-based" depolymerization process that breaks down polymers into monomers, allowing the polymers to be reclaimed, repolymerized, and converted back into raw materials such as fibers.

The duration of plastics in nature has been a significant concern for many environmental advocacy groups and activists, although these concerns are often misinformed, as reviewed in **Chapter I: Unfortunate Disposal** section.

Consequently, the news about innovations in plastic degradation can seem highly promising to the public. However, as the saying goes, the devil is in the details. Like many other seemingly wonderful technologies, this one may also fall short of its promises in many practical aspects.

Potential Greenwashing: One of many reasons this might be an instance of greenwashing is due to the lack of disclosure regarding the practical and commercial implementation of the technology. With the lack of the information, it may be a project designed more to enhance the company's image than to provide actual and meaningful environmental benefits, fitting the typical definition of greenwashing.

Implementation Concerns: When I came across the news article about this claim, I was skeptical because I understood how costly and time-consuming it would be to implement such a technology. An analogy can be drawn from food fermentation: producing a kilogram of cheese is much more expensive than pasteurizing a kilogram of milk obtained from a cow because it involves fermenting milk with food enzymes over a considerable amount of time.

This depolymerization-with-enzymes process would require vast land areas to store textile goods, inject a large quantity of useful enzymes, and wait for the depolymerization process to complete. Additionally, creating optimal conditions for enzymes to work, such as high temperature and humidity, would consume significant energy and inadvertently induce other potential harm such as carbon emissions for the environment.

Established Science: Moreover, the concept of this process is neither new nor noble. Laboratories have long created such conditions to study depolymerization, particularly in relation to the biodegradability of different forms of plastics, a topic covered earlier in **Chapter I: Section 3 Unfortunate Disposal**. Once depolymerized, it is a basic science of thermoplastics that it can be repolymerized.

Environmental Impact: From an engineering perspective, I am skeptical about the net environmental benefits of implementing this technology. The carbon footprint of the fibers produced by this method would be extremely high, attributed not only to creating the conditions for the elevated enzyme activities but also the byproducts of the associated chemical reactions. Additionally, there are other unknown environmental impacts about the industrial cultivation of enzymes.

Given the extremely high efficiency of the conventional synthetic fiber production, discussed in **Chapter II: Section 8. Synthetic Fibers**, this technology seems more like a consumer-appealing image enhancement than a genuinely sustainable solution.

Lack of Transparency: In this particular case, the company should have provided specific details about the production scale using this technology, the cost for consumers to purchase products made this way, and clear insights into the comprehensive environmental impacts.

***

Unfortunately, the news article I encountered announcing this program lacked such critical information. Without transparency and detailed disclosure, such initiatives risk being mere greenwashing rather than contributing to genuine environmental sustainability.

Case #4 - A fashionwear company claims to use natural fibers instead of synthetic ones for environmental sustainability. Organic fibers they use are supported by traceability certification.

When I engage in textile related discussions with those around me, two of the most surprising insights I share are: 1. Natural materials do not always outperform synthetic materials based on objective performance metrics, and 2. Natural materials are not inherently more sustainable than synthetic alternatives.

In this case study, I will highlight key points, some of which built on earlier discussions, that are often overlooked by the general public and environmental interest groups when evaluating the environmental sustainability of different fiber categories:

Performance Aspect

Many people believe that natural materials outperform synthetic ones, and this belief, among other factors, is often reflected in higher retail prices of textile goods on the market. For instance, winter coats made with high-quality down are typically much more expensive than those insulated with synthetic materials. However, scientific analysis shows that this perception of superior performance is situation-dependent, and in many cases, synthetic materials can actually perform better. For example, down's natural tendency to absorb moisture can reduce its ability to create an ideal formation of dry air pockets necessary for optimal insulation, leading to a lower level of warmth compared to their synthetic alternatives in humid weather conditions or when the humidity increases in the microclimate – more information about the microclimate presented in **1.4.1 Heat Reflective Technology**.

A crucial aspect of this performance discussion in an environmental context is that underperformance can lead to overconsumption*. For instance, someone wearing an expensive down coat might need to layer heavily with a base layer, woolen long-sleeve shirt, and thick sweater to stay warm. If the coat provided sufficient warmth on its own, fewer or lighter layers would be needed, ultimately benefiting the environment by reducing the overall textile consumption.

> * This topic of **"overconsumption driven by underperformance"** touches on many related factors, including weather conditions, material types, construction methods, and thermal technologies for various textile goods where warmth is essential. While there are some overlapping areas with environmental aspects in these discussions, contents are too large and extensive to be included in this book. Thus, I reserve a detailed exploration of these subjects for another book, expected for publication in the early part of 2025.

Environmental Aspect

As this subject has been extensively explored throughout the book, following is just a recap:

High Chemical Usage: Natural fibers involve significantly higher quantities of harmful chemical use throughout their entire lifecycles, from farming (with pesticides and fertilizers) to manufacturing, consumption, and disposal in comparison with synthetic fibers.

Duration and Bioaccumulation: Persistence of harmful substances in the environment and their tendency to accumulate in living organisms is a critical concern. Given the nature of fibers themselves and chemicals used, natural fibers contribute to more environmental and health impacts than their synthetic counterparts.

Water Consumption: Enormous quantity of water required for natural fibers spans from farming to manufacturing and consumption stages and exceed that required for synthetic ones.

Food Security Issues: Cultivation of natural fibers can impact food security, as agricultural land is diverted from food production whereas synthetic fibers do not create negative impacts.

Energy Consumption: Natural fibers invariably involve higher energy consumption from manufacturing to consumption stages compared with synthetic fibers.

All of the above points were part of the 16-point Lifecycle Assessment (LCA) analyses performed in **Chapter III**.

If a company claims that natural fibers are more sustainable than synthetic fibers without addressing these issues, it is, at best, misleading and, at worst, a clear example of greenwashing and an act of harming the environment.

The discussion around "organic" fibers and traceability certification programs introduces additional complexities.

These topics will be further explored in **Section 19 Loopholes in Certification Programs** later in this chapter.

Case #5 - A company specializing thermal insulation materials asserts that one of their synthetic product offerings is "biodegradable."

This is another typical example of certain companies and brands either exploiting the public's lack of knowledge on related subjects or failing to make sufficient efforts to understand their own claims, thereby engaging in greenwashing.

I presented the details of the potentially greenwashing aspects of such claim in **Chapter I: 3.3 Biodegradability** section. A particular greenwashing aspect of this claim is that thermal insulation materials involve significant structural challenges in their biodegradability. With the presence of enveloping fabrics, the materials are not exposed to direct contact with the natural elements of decomposition. Given the variety of fabric types used in a wide range of the goods containing thermal insulation materials, it is implausible that the company can predict the biodegradation of their materials.

Making such claims without addressing these concerns would be a typical act of greenwashing.

Case #6 - An environmental NGO issues a research paper and makes such claims as *"plastics have never been recyclable and never will be"*, *"plastic recycling is not technically or economically viable at scale"*. **Under the same themes, it blames petroleum companies for lying to consumers about the recyclability.**

As the public concern for the environmental sustainability has intensified, there has been an unfortunate trend of scapegoating certain industries and organizations, often leading to their vilification. A pertinent example is the petroleum industry. A few months before the time of this writing, I encountered a paper titled "*The Fraud of Plastic Recycling*", issued by an influential think-tank group based in Washington D.C. This paper, released in February 2024, lists a PhD holder as the primary author, with three co-authors and three additional contributors.

No Science: The report makes numerous claims that fundamentally contradict the established scientific principles. In my mind, it is perplexing how such erroneous assertions could be published especially when considering that Washington D.C. is home to the headquarters of many influential science and industry associations, including the American Chemical Society (ACS) and the American Petroleum Institute (API).

Science: As explained in **Chapter II: 8.1 Definition of Plastics** subsection, thermoplastics can be converted into a molten state using only heat energy. This principle has underpinned the flourishing plastic recycling industry for many years already. While I have previously expressed concerns about the conventional recycling methods due to their negative environmental impacts, the recyclability of the thermoplastics and the economic model of the recycling have been validated through the operations of numerous profit-seeking organizations.

Furthermore, many claims and arguments presented in the paper lack even the most basic level of scientific understanding. Such actions align with one of the extended definitions of greenwashing discussed earlier in this chapter:

f.   Misleading the public with misinformation

In conclusion, the need for rigorous scientific validation in sustainability claims cannot be overstated. Misleading the public with unsubstantiated assertions not only undermines genuine environmental efforts but also contributes to the unwarranted vilification of targeted industries.

Case #7 – EU Commission's campaign, "*It's time to put fast fashion out of fashion.*"

When I encountered this campaign, I was astounded. While I can vaguely comprehend the EU Commission's intention, rooted in a common perception that fast consumption leads to overconsumption, this act involves numerous aspects of greenwashing. The flawed insights of labeling certain textile goods as "fast fashion" and potentially negative environmental outcomes are detailed in **Chapter I: Section 2 "Fast Fashion"**.

Given that the EU Commission is one of the highest authorities globally, and a significant portion of the public places unwavering trust in their actions, the consequences of such a misdirected campaign could have long-term effects on the environment.

Despite this, I believe their intentions were good. Faced with public pressure to act on environmental issues, the EU Commission felt compelled to respond, resulting in the launch of this program. This situation highlights the crucial need for comprehensive education for the individuals involved in these organizations and environmental interest groups. Proper understanding and knowledge are essential to ensure that earnest efforts do not inadvertently cause harm, more of which will be discussed in **Chapter V: Education, Education and Education** section.

Section 17 Self-Proclaimed Expertise

If there are too many captains on the boat, it goes up the mountain. **– Korean proverb**

This Korean proverb aptly describes the current state of global sustainability and environmental management. To make it even more fitting to the situation, I slightly modify it as following:

If there are too many <unqualified> captains on the boat, they do nothing but <wreck> the boat.

Unfortunately, this sentiment reflects the reality of many initiatives and actions taken worldwide today in the name of sustainability and environmental management. Numerous individuals and organizations, despite their good intentions, are hindering global efforts. I recognize that this assertion may be provocative and raise eyebrows for many. However, it is an undeniable truth that there are too many "self-proclaimed experts", who significantly lack necessary scientific knowledge and relevant experiences.

Definition

My search for the definition of "self-proclaimed experts" in various dictionaries did not yield satisfactory results. Instead, I found an insightful definition, written in an open discussion forum on the internet by a person named Suvija J. V.

> "Being a self-proclaimed "fan" or "expert" in a subject while lacking knowledge about it usually indicates a person's enthusiasm or desire to be associated with that

particular topic. However, it may also suggest a level of overconfidence or misconception about their actual understanding. Claiming expertise without the necessary knowledge can be misleading. True expertise typically involves a comprehensive understanding, skill, and experience in a specific field. Overstating one's knowledge may lead to inaccurate assessments or advice." - Suvija J V

Fertile Ground for Self-Proclaimed Experts

Sustainability and ESG have become a field where individuals can readily claim their expertise. As these topics are among the most pressing issues globally, people from diverse personal and professional backgrounds participate in related activities. For example, many professionals from the media industry have taken prominent roles in sustainability and ESG efforts worldwide. Additionally, individuals with varying degrees of experience across different sectors, including textiles, are contributing to significant societal efforts toward sustainability.

While professionals with a background in the textile industry might possess a greater familiarity with relevant topics compared to those from other fields, it is important to note that individuals in administrative roles or even those working in engineering positions away from manufacturing environment may not always have sufficiently accurate insights into the real-world impacts of their work.

Educated by Internet: The internet is particularly effective in the rapid spread of information, especially in English-speaking contexts. A notable example is the term "fast fashion", popularized by a New York Times reporter. The New York Times, a highly respected media brand, significantly influences global discourse. The widespread adoption of the term to the extent that it is recognized globally illustrates the effectiveness of its propagation, irrespective of the accuracy of the information.

Real-World Experts: In contrast, many experts with deep understanding of the real world of textile manufacturing and chemical applications in distant, non-English speaking countries are either silent or lack a significant internet presence. For instance, my observation, in a real-world manufacturing environment during my first job in Korea close to 30 years ago, of not seeing any rat in the dyeing department, indicating the toxicity of the area, while they were plentiful in other departments, allowed me to gain both practical and impactful insights of the toxicity of dyeing chemicals in general. Even if the internet was readily available at the time, my limited language skill in English would not have prompted me to propagate such information.

While it took 30 years for me to share this information with readers, the term "fast fashion" gained widespread popularity almost instantly, highlighting how the reach of the information sources, rather than the accuracy of the contents, can significantly influence public perception.

Interdisciplinary Experts: It is important to acknowledge that individuals can become true experts in the fields outside their educational and professional backgrounds through dedicated effort. For example, Ed Conway, a journalist, the author of "Material World," demonstrating deep knowledge and expertise in materials science despite his background in journalism, gaining my respect in such interdisciplinary expertise. In my view, his exceptional analytical skills and scientific aptitude allowed him to produce a masterful work that helps readers understand the impacts of key materials on our lives. However, such cases are rare, especially in the realm of the textile sustainability.

Absence of Presence: A major issue with this absence of the presence of the real-world experts in our global sustainability efforts creates increasing reliance on "internet-based self-education" by these "notion-based" experts. Their understanding is often shaped by the popularity and accessibility of information, rather than its accuracy and scientific validity. This discussion further leads to a deeper level of examination on the roles of internet search engine operators, which will be covered more in depth in **Chapter V: 21.2 Responsibilities of Search Engine Operators**.

***

As we undertake truly substantial challenges in the global environmental management, it is essential to have broad participations from the members of our global society. Each person may bear important roles in this effort.

In the meantime, the discussions of this section underscore the importance of the genuine expertise in the global sustainability efforts. While we appreciate the enthusiasm of numerous environmental advocacy groups and activists, wanting to take leading roles, there has to be a clear role-taking and leaderships in driving concerted sustainability efforts globally.

In this context, it is crucial that genuine experts whose knowledge and experiences extend to comprehensive areas of sustainability set impactful objectives and efficient global actions. Continually allowing self-proclaimed experts to drive global sustainability efforts can result in the continued inefficiency, risking ourselves as the bystanders of those, shooting darts blindfolded.

Section 18 Organisations Lacking Expertise

With the public's keen interest in the sustainability, it appears that some of the largest NGOs are environmental groups these days. Additionally, governments and their affiliates have been heavily involved in a wide range of environmental activities.

However, it seems that even some of the largest organizations are run or heavily influenced by "self-proclaimed experts". This issue is exacerbated by these organizations' lack of commitment to seeking accurate, unbiased, and science-based information. Consequently, many sustainability claims, actions, and movements driven by such organizations may inadvertently harm the environment.

In this section, I analyze two groups: The first group comprises textile companies and the second pertains to other stakeholders such as government organizations, their affiliates, NGOs, and activists.

18.1 Textile Companies

The examples of potential greenwashing cases discussed earlier highlight widespread misconduct even by some of the largest companies and brands in the global textile industry.

Responsibility: For a company to be responsible for its claims, it must invest sufficient resources to fully understand and transparently disclose all relevant information, a concept discussed earlier in this chapter. This includes details about source materials, processing methods, known environmental side effects and potential areas of harm.

Given that the actions of textile companies have direct impacts on the environmental sustainability, and our lifestyles, these companies must take their responsibilities seriously, perhaps more so than any other stakeholders.

Wrongful Accusations: While much evidence of corporate misconduct and greenwashing exists in the textile industry, there is also another side to the story. Occasionally, companies are wrongly accused by environmental interest groups. Some of these groups and activists can be overly aggressive, operating on misunderstanding and lack of expertise, sometimes resembling a form of modern "witch-hunting". Driven by anger, they seek to influence and mobilize others, often targeting textile organizations with their frustrations.

By their very nature, organizations are subject to public scrutiny, whether justified or not. Earlier, I discussed an example where the petroleum industry was accused of dishonesty due to a significant misunderstanding of the related scientific facts. Similar situations are increasingly emerging in the textile industry, often resulting in legal actions. These incidents are becoming more common reported in the news media around the globe, reflecting a complex relationship between public perception and industries.

Serious Disconnect: As I outlined some of the challenges that textile companies face in fulfilling their sustainability duties, addressing these issues can be a daunting task, particularly due to the complex circumstances surrounding the industry structure. For instance, a serious disconnect exists between the administrative functions and manufacturing operations of textile companies, often leading to misguided actions in their sustainability efforts. This gap stems from the historical development of the textile industry, particularly the migration of manufacturing, and has become deeply ingrained, unnoticed by even some of the most experienced executives in the industry. Following discussions explain such phenomena.

Economic Development and Textile Industry

Throughout the economic development of a large number of countries around the globe, the textile industry often plays a crucial role in the early stages of industrialization. This pattern was evident in England during the Industrial Revolution in the 18th century. The invention of the steam engine allowed for the construction of large factories away from river streams, which had been previously essential to be located close to them for utilizing waterpower. The flexibility and efficiency of the steam power led to rapid industrial output increases and accelerated wealth accumulation. As textiles are essential to human living and represent a significant improvement in lifestyle, the steam engine and energy revolution immediately benefited the textile industrialization.

Labour Requirements and Cost: The textile industry is characterized by its high demand for manpower. Large textile factories can employ thousands of workers. Although machinery automation has reduced the need for manual labor over time, and machines can be used for certain processes such as packaging and inventory management, the industry still relies heavily on skilled manpower in such roles as sewing due to the complexity and delicacy involved in those processes.

Additionally, labor costs in many underdeveloped and developing countries are low enough that investing in expensive machines to replace human workers is often not economically advantageous for many players in the industry.

Global Migrations: This economic reality has driven the global migration of textile manufacturing to regions offering lower labor costs. Ironically, the wealth amassed by a society, significantly contributed to by the textile industry in many countries, leads to an increase in labor costs. Consequently, the cost of manufacturing textile goods increases proportionally. While industries producing high-value-added goods such as heavy machinery, IT, etc., can accommodate such increases, the textile industry, with its traditional low-margin structures, cannot - In a large portion of textile manufacturing, the labor constitutes one of the most significant cost bases.

Throughout the history of the textile manufacturing migrations, major trends have emerged. Initially, the shifts occurred from UK and US, pioneers of the industrial revolutions, to Japan and other European countries. After World War II, there was a significant shift to parts of Asia. Then, another major shift occurred when China joined the WTO, marking the most significant transition in recent history. As China's economy developed and labor costs increased, coastal cities like Shanghai and Shenzhen could no longer offer competitive labor costs for the industry. Consequently, many factories in China have migrated either inland, where labor is still affordable, or to other countries. This shift has prompted the emergence of Vietnam, Cambodia, Thailand, etc., in Asia and some South American countries as new textile manufacturing hubs. The latest trend in migration is toward African countries, where manufacturing has not traditionally been a stronghold.

Structural Changes of Textile Industry

As a result of these manufacturing migrations, many Western textile brands operate in two distinct structural segments: the manufacturing in overseas facilities and the administrative operations in their head office locations. A large textile brand, for instance, may house its executives, designers, purchasers, engineers, marketers, HR, warehouse personnel, etc., in its head office, while maintaining its operational structures to work efficiently with their manufacturing partners in distant countries across borders and oceans.

Consequently, today's staff working in their head offices may have never had any opportunity to gain extensive and practical knowledge of real-world manufacturing.

Serious Knowledge Gap and Implications: Several decades ago, when textile manufacturing was still robust in many Western countries, industry experts often had a well-rounded understanding of both the manufacturing and administrative aspects of their businesses. This was largely due to the close integration of their operations, which naturally facilitated the exchange of knowledge and experience between professionals in a wide range of areas. However, those experts have long since retired, and over the ensuing decades, a significant knowledge gap has developed between the manufacturing and administration, a gap that continues to widen. Unfortunately, this disconnect contributes to a growing divide between "theoretically sustainable practices" and their actual impact in the real world.

Erroneous Efforts: For example, if a marketing team at a large brand company pursues a sustainability program using recycled fibers without ever setting foot inside a recycling factory, or even recognizing the necessity of doing so, they may automatically resort to a third-party certification program.

While the executives of the company might be satisfied with such actions, which require minimal resources and effort, simply relying on a third-party certification program can have significant implications for the environmental sustainability, as discussed earlier in this chapter.

Need for In-House Expertise

One of my main aspirations in writing this book is to encourage textile companies to develop their own expertise in the business areas related to ESG. It is not overly challenging, for example, to send an engineer to a "certified" recycling facility to assess potentially harmful impacts behind the certifications, or to assign a person to analyze "biodegradability" claims and understand the associated loopholes. Relying solely on third-party certification programs places a company at the mercy of external resources, which often reveal incomplete, improper, inaccurate, or even contradictory nature of their work.

By developing in-house expertise, companies can better navigate the complexities of their sustainability claims and actions. This responsible act will contribute positively to their environmental efforts and protect themselves from the pitfalls of greenwashing and undue criticism.

***

While several essential needs for textile companies have been identified, implementing these solutions is likely to be more challenging than it may seem, as achieving meaningful results will require earnest efforts from a wide range of stakeholders. Due to the importance of this issue, I will expand on this discussion in **Chapter V: Solutions**.

18.2 NGOs, Policy Makers, Government Organizations and Affiliates

In general, we accept authorities and their actions to lead our societies. For example, different political systems such as democratic, socialist and communist governments profoundly influence our lifestyles and even mentalities. The laws established by different governmental systems serve as the fundamental building blocks of social order. With the trust we place in these authorities, we generally rely on them to guide our societies in the right direction. Conversely, social values are typically shaped by the needs and wants of the public.

Disappointment: In recent years, a focus on the environmental sustainability has been highlighted as a strong social value by the public and we expect strong leadership from the authorities. However, I am profoundly disappointed with the actions governments and their affiliates worldwide have taken in this area.

Earlier in this chapter, I discussed the EU Commission's campaign, "*It is time to put fast-fashion out of fashion*", as an example of greenwashing. Instead of targeting unidentified entities with vague notions of "fast fashion" out of lack of understanding on the mechanics and dynamics of the textile industry, the Commission could have adopted the following sentiment in its attempt to curtail the textile consumption rate of its citizens:

"The citizens in the developed world have plateaued their consumption of textile goods for the first time in the human history. This is commendable. However, for the sake of the environment, we need to do better. Overall consumption is still growing globally, as underdeveloped and developing countries consume more followed by rising income levels. This trend will continue until they reach similar income levels as those of the developed world. Therefore, let's reduce our own consumption and try to help others and our environment together."

The missteps of these organizations underscore the importance of developing their own expertise sufficiently. This discussion also warrants a further exploration and will be presented more in depth in **Chapter V: Solutions**.

Section 19 Loopholes in Certification Programs

Realistically, certification programs in environmental and sustainability sectors should provide comprehensive frameworks that align with the global objectives for environmental protection and sustainability. Certifications should instill confidence, even in those without expert knowledge, that the certified subject matter has a positive global environmental impact. Unfortunately, many certification programs fall short of this standard.

Failure: As of this writing, one of the most recognized certification programs for recycled fibers overlooks critical issues such as the generation of microplastics during recycling and manufacturing processes, the reduced strength of recycled fibers and its impact on consumption, and the increased release of chemicals into the ecosystem. Despite these negative impacts, fibers are still awarded certification as long as they are recycled, and products are branded with logos, often in green tones, designed to convey environmental sustainability to consumers. Textile brands can display these logos on their merchandise as long as a certain threshold of the recycled content is met.

Even some of the most reputable companies in the global textile market participate in this practice, which contributes to significant misperceptions about the actual environmental impacts of these certified fibers.

Many Failures: This issue is not isolated, as it is just one of numerous examples of the loopholes of certification programs. Many of them cover only limited scopes and omit other significantly important aspects of sustainability. Despite this, they are accepted, or even demanded, by the public, and sometimes endorsed by governments and NGOs, due to a lack of knowledge or effort to seek better understanding.

"Chicken or Egg" Dilemma: There may be a debate akin to the "chicken or egg" dilemma in this context. Textile companies might argue that the public demand drives their actions, while the public might assume that the sustainability claims made by these companies are the results of thorough work in comprehensive environmental scope. In my opinion, the responsibility lies primarily with the companies and lack of governance of the authorities of their jurisdiction. Furthermore, the public generally places too much trust in large corporations, which, based on my decades of experience in the corporate world, often do not deserve such high levels of trust.

Commercial or Operational Benefits: Many organizations offering certification programs operate on profit-driven models and are privately owned. These programs can be extremely expensive, costing tens of thousands of dollars for initial certification and thousands more annually for maintenance. Consequently, even if a startup company develops truly sustainable products and technologies, these programs are often financially inaccessible, limiting consumers' choices. In my view, such phenomenon is a form of industry exploitation in making the programs as a "money game" and limiting the public's accessibility to their advantages. In other cases, non-profit based organizations may be established with elevated cost structures, prohibiting them to offer more affordable prices.

Nations Unified for Environment and Sustainability (NUES): In my view, certification programs should be governed by a global organization dedicated to sustainability, free from political and commercial influences - I will present this concept with more details in **Chapter V: Solutions** with my views on the roles played by the existing authorities such as United Nations and their failure on even some of the most critical areas of the global environmental and sustainable management.

Information provided by certification programs must include not only advantages but also disadvantages, concerns, and impacts, which can potentially create issues for the environment. This approach would enable consumers to gain more balanced information and make informed judgments about their true environmental benefits. Conversely, it would prevent consumers from simply relying on a green logo without understanding related context.

In the meantime, I urge all stakeholders, including the organizations that govern certification programs, their corporate customers, and government organizations that provide financial or other forms of support, to invest in developing their own expertise in comprehensively understanding the true environmental impacts of their programs. This investment is crucial to leading the global society in the right direction.

Chapter IV Post-Chapter Commentary

In this chapter, I presented several examples of greenwashing. The astonishing nature of the phenomena encompasses several key factors:

Sheer Number: It is truly astounding to witness the sheer number of obvious or highly potential greenwashing claims. While I presented a handful of examples, there are numerous others.

Rudimentary Nature: A large majority of these greenwashing claims is made without even the most basic ground for accurate science.

Authorities: Organizations of all sizes, including some of the largest NGOs, private organizations, and even governments, are engaged in a broad scope of greenwashing activities.

Recognizing the depth of the issues, I shared my views on several key causes of the undesirable situations in the state of the global environmental management and efforts. Accurately understanding the root causes is a critical step towards finding solutions. While many other causes can be attributed to, the main trend of the situation may be formed by the following three key elements:

Self-Proclaimed Experts, generating misinformation founded upon notions rather than science.

Widespread Dissemination of such misinformation, influencing the public.

Organizations, lacking efforts to develop their own expertise, resulting in misleading the public or being easily influenced by self-proclaimed experts.

Examining these causes and understanding their implications lead naturally to the discussions of the next chapter: **Solutions**.

Chapter V

****

Solutions

Pre-Chapter Commentary

Quote of the chapter

"Be transparent! The nature of political systems often leads governments to highlight successes and downplay failures. Significant social and financial resources are spent on promoting positive outcomes through advertisements, funded studies, etc. We generally accept it as a social component of governance. However, in the perspective of environmental management, it is unrealistic to expect actions that only yield positive outcomes without any adverse effects"

***

The title of this chapter, "Solutions", places upon me the significant burden of presenting my views and suggestions as we are in a serious state in the environmental management for our sustainability. The issues are of profound nature and critical for the future of our existence.

Expanding the Scope Beyond Textiles: While my focus thus far has been on the textiles, this chapter introduces broader perspectives to highlight that environmental issues cannot be effectively addressed without a comprehensive global view that considers all possible angles. Contributing factors influencing the outcome of our environmental efforts are interconnected through shared environmental elements such as water and air streams. This interconnectedness underscores the need for holistic solutions.

Consensus and Genuine Intentions: There is a general consensus on the urgent actions required to protect our environment and ensure sustainability. Most environmental interest groups genuinely aim to benefit the environment and humanity. With the collective efforts by the global society, guided by science-based directives, overcoming seemingly insurmountable hurdles is ahead of us. In this chapter, I present my views and suggestions on achieving such goals.

Section 20   United Efforts

We encounter numerous discrepancies in our personal, professional, and societal lives. Some of the topics discussed in the earlier chapters exemplify significant discrepancies currently occurring worldwide in the realm of the environmental sustainability. One striking example is the practice of affluent countries shipping large quantities of textile waste, including heavy winter goods, to underdeveloped and developing countries in warmer climates that do not experience winter. These items are often burned or buried in the destined countries without any environmental control, primarily because environmental regulations and monitoring in these countries are less stringent or non-existent. Similarly, the prevalence of greenwashing, committed even by some of the most reputable and respected organizations, highlight the lack of effective efforts in our global community.

Lacking Rationale: The gravity of our concerns on such discrepancies stems from the rudimentary nature of these issues. For instance, if we consider that contaminated water and air from burial and incineration sites in Africa can travel and impact other regions around the globe within days, weeks, or months, it becomes clear that shipping waste across international oceans, thereby creating substantial carbon footprints, is a nonsensical act. This practice, lacking even the most basic communal and science-based rationale, has no place in the sustainability efforts and also, constitutes a form of humanitarian cruelty.

In this section, I invite readers to join me in exploring many other astonishingly inefficient and poorly conceived efforts that surround us in non-textile areas. Following this examination, I will present my perspectives on more concerted and effective strategies for addressing these critical issues under the context of the global nature of our efforts.

20.1 Poorly Made Efforts – 1: UN case

In 2008, the United Nations Educational, Scientific and Cultural Organization (UNESCO), an agency of the United Nations (UN), designated the lagoons of New Caledonia in the South Pacific Ocean as a world heritage site, recognizing their ecological and cultural significance. Paradoxically, another UN body, the International Seabed Authority (ISA), granted permissions for mining activities in the areas overlapping or adjacent to this heritage site to explore for metals and minerals such as copper, nickel, cobalt, etc.

This situation raises significant questions about the coordination and efficacy of global efforts made by various organizations, including the UN, an organization considered as one of the highest authorities in the global governance. How could such a conflicting decision be made? While there are inherent complexities in balancing environmental protection with resource development, this scenario exemplifies inefficiencies and contradictions that often plague environmental management efforts worldwide.

Historical Context of Undersea Resource Exploitation: The potential of undersea resources has been recognized for over a century. The first undersea drilling for petroleum occurred in 1896 off the coast of California. The wealth of precious metals such as gold, silver, copper, cobalt, etc., in the seabed, essential for such industries as electric vehicles (EV), information technology (IT), etc., has become increasingly apparent in recent years. Given the purity and higher densities of these essential metals compared to their terrestrial counterparts, their excavation offers significant economic advantages.

However, heightened awareness of the environmental impacts of mining activities has prompted significant global debates. Many people may be unaware of the extensive and often devastating consequences of thousands of years of mining activities by humans. Industrial efficiencies have accelerated these impacts, leading to a significant environmental toxicity creation. While I will expand further on this subject in the next subject, this discrepancy between the two UN bodies highlights the nature of poorly coordinated effort.

20.2 Poorly Made Efforts – 2: Electric Vehicle (EV) Conversion

It may surprise many readers that I chose the EV conversion as a topic in the context of "discrepancies" in our environmental efforts. Before delving into the analysis of the global EV conversion initiatives, I wish to clarify that this discussion is not intended to determine whether pursuing EV conversion is right or wrong. The global shift towards EV is widely recognized by the public as a crucial step in reducing carbon emissions and addressing climate changes.

My concern lies in many unknowns surrounding the true environmental impacts of this conversion. As we delve deeper into this discussion, readers will begin to understand the complexities involved in the transition to EVs and recognize the significant gaps in our knowledge regarding critical environmental factors. In fact, this lack of understanding is a discrepancy in itself. As global citizens, we deserve a comprehensive understanding of the broader implications, beyond the carbon footprint considerations, of investing trillions of dollars into initiatives that could potentially result in significant environmental side effects. Moreover, based on my analysis, the anticipated benefits in terms of the carbon emission reductions may be overstated, an issue that will be explored in greater depth in this subsection.

I understand that some readers may have already formed skepticism about the statements I made above and might question the integrity of the subsequent discussions, perceiving them as biased or ill-intentioned. However, I encourage you to approach the contents with open mind and read through the entire discussion before forming a final judgment. As in the previous chapters, you will find that the issues at hand are far more intricate and complex than the simplistic notion of "*no emission equals better for the environment*".

While some of the brief statements in this subsection, particularly at the outset, may challenge your existing beliefs, I provide ample information and context throughout the relevant topics. This approach is designed to first introduce key concepts succinctly, and then progressively expand the discussions with more detailed scientific analysis. This method will help readers gradually become more familiar with the underlying science and reasoning that support these perspectives.

Important Discrepancies

Although the EV conversion has become one of the most significant efforts of our global societies for the environmental sustainability, receiving almost unanimous support from the majority of the public, analytical minds have raised several important questions in various aspects.

A simple question that has persisted in recent years is, "*Do we even know if there is a net benefit for the environment from the massive EV conversion, factoring in the positive and negative impacts?*" This general question is often met with even simpler assertions, such as "*benefits outweigh negatives*". The issue in more scientific views in closer examinations is that there is insufficient evidence in the overall environmental impacts for both "benefits" and "negatives".

Legitimate Concerns: Several more specific concerns have emerged. One of them is the toxicity concern from the massive increase in mining activities for the metals and other forms of minerals essential for the EV batteries, such as lithium, cobalt, nickel, Rare Earth Elements (REE), etc. – An extensive exploration on the toxicity aspect from the mining activities will be presented in **C. Toxicity Factor** later in this discussion.

Coming closer to a more conventional realm of the public discussions for the conversion; the environmental benefit of EVs is contingent on the source of the electricity used for charging. In the regions where the electricity is predominantly generated from fossil fuels, the net reduction in carbon emissions may be far different from the general perception. Later in this discussion, I will make an attempt to illustrate the current state of the carbon impacts with some statistics available in the public domain.

Large Mistakes of Ignorance and Downplaying: Despite these valid concerns, the momentum towards the global EV adoption appears unstoppable, often sidelining dissenting voices. This overwhelming push for EVs raises a serious question of whether alternative perspectives are being adequately considered in the global governance steering the movement.

Continuing to ignore such concerns would be a large mistake, especially given the extremely poor track records of human judgements and decisions, significantly harming our environment. This topic will also be further explored with specific examples in subsequent discussions.

***

With these introductory remarks, I now invite readers to join me in exploring these topics, identifying areas of concern and unknowns. Even advocates of the EV transition may find value in gaining a more comprehensive view in the areas that they may not have previously considered. I encourage all readers to, once again, keep an open mind and refrain from dismissing these concerns, no matter how minor they may appear, unless they have clear, convincing data to address them. The potential impacts of even seemingly small issues may be far more significant than they might presume.

20.2.1 Environmental Footprint Analysis

Claimed Benefits: Stakeholder and Proponents of the EV adoption often use various metrics to demonstrate its environmental benefits. For instance, a 2020 study in Nature Communications* suggested that widespread EV adoption could reduce particulate matter (PM) emissions of 2.5 micrometers (PM2.5) or smaller by 50% or more in urban areas where coal-fired power plants are not the primary electricity source.

> * Citation - Buonocore, J.J., Nguyen, T., Driscoll, M.A. et al. (2020). Article 5222.

At first glance, this figure appears impressive and suggests a clear environmental advantage.

Scientific Analysis: However, a deeper examination reveals a more complex picture. There are several key premises to consider when analyzing this claim.

First, the particulate matters come in different sizes, with both PM2.5 and PM10 (10 micrometers or smaller) being a significant contributor to air pollution. These particles are small enough to enter the body and affect organs. Therefore, accounting for only PM2.5 and evaluating overall improvement in the air quality is erroneous. While some argue that the tailpipe emissions relate more to smaller sizes of PM such as 2.5 and that PM10 comes from the non-exhaust sources such as tire friction against road surfaces, break wear, etc., larger particles remain and continually breakdown just like the smaller ones, persisting in the environment.

Second, studies suggest that the non-exhaust emissions from tire and brake wear, as well as road surface abrasion, far exceed those from the tailpipes. With the implementation of strict emissions regulations, modern vehicles produce much cleaner exhaust emissions. In fact, some studies estimate that non-exhaust sources account for 50% to 90% of the PM10 and PM2.5 emissions of the total vehicle pollutants, with some claiming these emissions are up to 1,000 times higher than the exhaust emissions.

Third, the non-exhaust sources of PM contain harmful chemicals which are different from those from the tailpipes. For instance, 6PPD, a chemical used in tires as an antioxidant and antiozonant, is known to produce a byproduct (6PPD-quinone) that is highly toxic to humans and wildlife.

Fourth, Identifying and measuring the different sources of PM is extremely challenging in the real-world conditions. Airborne particles from various sources mix and travel great distances, making it difficult to attribute them accurately.

While the study focuses on the air quality and their associated health improvements from the fleet electrification in the U.S., other cases, such as those in China and UK, offer broader perspectives and help gain practical knowledge that are relevant to this discussion.

London, U.K 150 Years ago: In the late 19th century in London, England, there was a significant fashion trend where people wore black clothing. Light-colored garments would darken quickly from soot and particulates in the air, even though there were far fewer vehicles on the road at the time. This highlights that the non-vehicular sources were the primary contributors to the air pollution in that era, reinforcing the idea that the non-exhaust sources can be significant.

Fig.53 London Air Pollution – Late 19th Century

https://theconversation.com/air-pollution-in-victorian-era-britain-its-effects-on-health-now-revealed-87208

China Today: China, which began its large-scale EV conversion efforts earlier than most countries, offers a useful case study. As of this writing, reports indicate that more than half of the new vehicles sold in China are electric or hybrid, and the cumulative number of these two vehicle types accounting for an estimated 10 ~ 12% of all vehicles on the road. This contrasts with the figures in the United States, where approximately 18% of the new vehicle sales are EVs and only 1 ~ 2% of the total vehicles on the road are electric.

Large quantities of PM2.5, PM10 and even larger sizes are produced by coal power plants and industrial activities. Estimates indicate that coal power plants account for as much as 50 ~ 60% of PM2.5 in heavily industrialized and urban areas, particularly before stricter environmental regulations and cleaner technologies had been implemented. Similarly, other industrial activities such as manufacturing, steel production, and cement production are estimated to contribute approximately 40 ~ 60% of the total PM2.5 emissions in China. Combining these figures suggests that these two sources of the PM2.5 emissions could exceed 100%, which is clearly impossible, while the vehicle emissions are not even included. Although accurately estimating different sources of the particulate matters in air poses significant challenges, if only feasible, to achieve a high degree of accuracy, one clear takeaway from these considerations is that coal power plants and other industrial sources are a significant contributing factor in air pollution.

***

The case studies from the UK and China provide valuable insights into interpreting the claimed benefits of the EV adoption. Although the Nature Communications study specifies conditions like *"urban centers where coal-fired power plants are not the primary electricity source"*, it is implausible to think that there exists an urban center where air is enclosed in a bubble where estimating different sources of PM could be made with a high degree of accuracy. Air flows and carries pollutants far and wide. In fact, there have been ongoing debates among some Asian countries, where neighboring nations claim that the air pollution from China travels across borders and even oceans and impacts the air quality in their countries. Another example was presented in **Chapter II: 10. Manufacturing** where the air quality of Uzbekistan did not correspond to the high level of soil and water pollution in the environmental state of the country, indicating the nature of air circulations spreading pollutants around.

Considering these complexities, the claim that the EV adoption could lead to a 50% or greater reduction in the PM2.5 appears considerably exaggerated. In fact, it can be outright wrong as some studies show that EVs contribute to increased particulate matter pollution due to their greater weight, as will be discussed further in **B. Material Demand** later in this subsection.

While we recognize that this Nature Communications example comes from a specific study with a limited scope, regardless of its intended purpose or the accuracy of the claimed effects, it highlights the importance of critically assessing such claims and the public's common perceptions of the EV benefits. Blindly accepting these claims benefits only the stakeholders promoting the conversion, while potentially causing harm to the environment.

With this perspective in mind, I aim to present the related information in this discussion based on sound science, while acknowledging that there is limited information available in certain areas. It starts with the carbon footprint impact analysis as the most highlighted benefit of the conversion by the proponents.

A. Carbon Factor

There are two parts of the caron emissions we must consider: 1. The emissions related to the power generation to satisfy EV's electricity needs and those of producing fuel for Internal Combustion Engine (ICE) vehicles, and 2. The industrial activities related to the EV production as well as the supporting infrastructure construction and maintaining the energy supply channel of ICEs.

Higher Material Demand: The shift to EV involves a significant overall increase in the use of related materials, which contribute to increased industrial activities and their associated carbon impacts.

For instance, EVs typically require two to five times more copper than traditional ICE vehicles. Conversely, renewable energy sources, such as solar panels and wind turbines, an essential element for the environmental benefits of EVs, also demand substantial amounts of copper. According to a projection from Goldman Sachs, the demand for copper related to green technologies is expected to increase by 5.5 times between 2022 and 2030.

https://www.mining.com/goldman-doubles-down-record-high-copper-price-within-a-year/

Carbon- and Greenhouse Gas-Intensive Nature of Mining and Refining: Mining and refining copper as well as many other minerals used in EVs is highly carbon- and greenhouse gas-intensive. For example, the processes involved in the copper mining, from ore extraction to crushing, grinding, flotation concentration, high-temperature smelting, and refining, generate significant quantities of CO_2, CH_4, N_2O, SF_6 and fluorinated Gases (HFCs, PFCs, CFCs). Moreover, these processes also require substantial amounts of water for diluting and separating the metal through a floatation method. Including the energy required to transport water from sources to processing facilities and treat it before releasing into the environment, the whole process demands intensive use of energy and chemicals, which in turn creates significant amounts of potent greenhouse gases.

Copper is just one example. Many other minerals essential for the EV batteries, such as lithium, cobalt, manganese, graphite, Rare Earth Elements (REE), etc., undergo similar mining and refining processes.

Electricity Sources: A significant portion of the electricity used to power EVs is generated from traditional fossil fuel sources, such as oil, natural gas, and coal. As of 2021, approximately 60% ~ 65% of the world's energy production was derived from fossil fuel. Alongside hydropower, one of the oldest energy sources, these traditional methods account for 75% ~ 80% of the global electricity production.

Solar and Wind Energy: Solar and wind contributed about 14.3% of the world's electricity in 2022, according the ChatGPT. It is important to note that the estimates of this nature can vary. For instance, a Google search might suggest the contribution of the solar and wind in the range of 10% ~ 13% for the same year. Therefore, these figures should be interpreted with caution, acknowledging a certain degree of variability.

Using the higher estimate of 14.3%, it means that the maximum contribution of the solar and wind energy to the electricity used by EVs in 2022 would be 14.3% in the power generation perspective, with the remainder coming from the traditional sources. While 14.3% is a significant figure, it is crucial to examine whether this 14.3% represents the net benefit of the global carbon footprint.

Since EVs currently fall predominantly within the passenger vehicle category, converting the figure to reflect the proportion of the carbon emissions generated by the passenger vehicles provides a clearer picture of their true carbon impact.

I once again quote ChatGPT:

Passenger vehicles are responsible for approximately 15-20% of total global carbon dioxide (CO_2) emissions from all sources. This includes cars, SUVs, and light trucks used for personal transportation.

Carbon Impact Analysis for Power Generation: From the above figures, we can calculate the overall carbon footprint impact from EVs is in the range of **2.0%** (=0.143x0.15) ~ **2.8%** (=0.143x0.2).

Then, we can reflect that the Energy is 73.2% of the total greenhouse gas emissions as demonstrated in ***Fig.54***, which results in **1.5%** ~ **2%** (=0.02 x 0.732 ~ 0.028 x 0.732) actual contribution from EV.

Fig.54 Global Greenhouse Gas Emissions by Sector

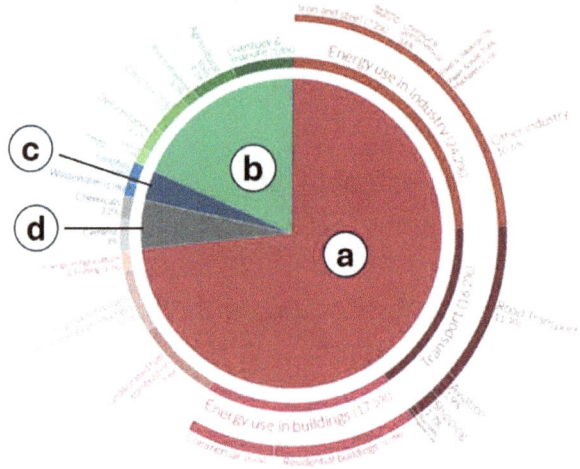

(a) – Energy **73.2%**, (b) – Agriculture, Forestry & Land use **18.4%**, (c) – Waste **3.2%**, (d) – Industry **5.2%**

https://ourworldindata.org/ghg-emissions-by-sector

The calculated figures of the carbon impact in the range of 1.5 ~ 2% may be astonishing for many. While I consistently apply caution when interpreting statistics in general, this analysis clearly serves its purpose for the legitimacy of such concerning voices as: "*Does this level of benefit justify significant environmental side effects it creates?*" and "*Is the massive investment, including hundreds of billions of dollars in subsidies using the public funds for lucrative businesses and affluent Western consumers, truly worth it?*", "*Could similar or better results be achieved through other means without incurring detrimental side-effects?*".

Inherent Complexity: Calculating the true carbon impact of EV is extremely complex. Due to this complexity, many figures available on the internet and other sources are invariably driven by simplistic approaches.

For instance, ICE is reported to create approximately 200g of carbon on average per kilometer driven. In case of EV, it is reported of approximately 150g for the same distance driven if the electricity is supplied by coal power plant. Although many people, including myself, may be surprised to see a difference of only 25%, this is not the whole picture.

A.1 Factors involving ICE Energy Creation: The *Fig.55* illustrates the process in which the energy required for ICE is created.

Fig.55 Energy Conversion Process – Fossil Fuel

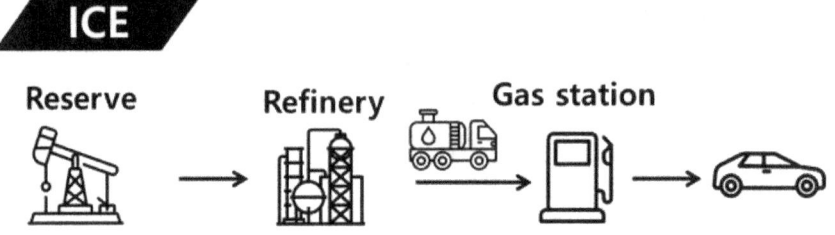

Factors related to the environmental impacts are:

Existing Infrastructure: Much of the infrastructure for oil extraction, refinery, operational systems for transportation and gas stations already exists. The petroleum output quantity often becomes a hot geopolitical agenda around the globe as, for instance, Organization of the Petroleum Exporting Countries (OPEC) decides to reduce or increase their oil production outputs based on economic or political considerations. This situation proves a considerable surplus in the existing capacity to be able to accommodate demand fluctuations.

Operational Efficiency: The petroleum industry has evolved over the course of two centuries to achieve a remarkable overall efficiency, driven by the profit-oriented nature of businesses striving to minimize costs. This efficiency is evident in the fact that petroleum companies have operated independently, in contrast to EV companies which have received hundreds of billions of dollars in subsidies from the public funds of many countries worldwide.

Lack of Mineral Mining: One of the most significant benefits, which aligns closely with the central theme of this book and is crucial for the environmental sustainability, is that it eliminates the need for mineral mining. This advantage will become more apparent in the subsequent discussions, where we will explore in-depth the environmental damage caused by the mining industry.

A.2 Factors involving EV energy creation: *Fig.56*

illustrates the process in which the energy required for EV is created.

Fig.56 Energy Conversion Process – EV, 2022

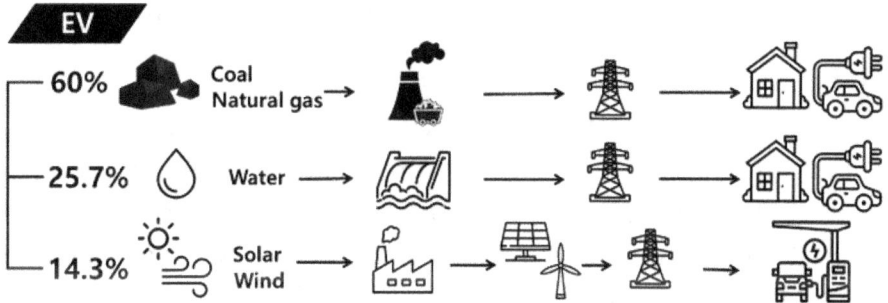

Compared with the simpler supply chain structure of ICE, EV requires added level of industrial activities. Factors related to the environmental impacts are:

New Infrastructure: A large portion of the required infrastructure must be built anew. For instance, the global demand for solar panels and wind turbines have created significant industrial activities including mining, manufacturing, transportation, construction, etc. Producing silicon, requiring thousands of °C or °F in industrial furnace for melting sand, and mining metals to wire solar panels and wind turbines, transporting them over large distances, building battery factories with gigawatts of capacities and power grid infrastructure, etc., are just a small sample of the requirements of the new resources, accompanying significant environmental impacts.

Manpower and Operations: Constructing new infrastructure for a substantial output requires significant amount of manpower, extensive operational systems and material demands while generating a considerable quantity of carbon and other greenhouse gas emissions. While these projects can benefit local communities by boosting their economy, they are not advantageous from an environmental perspective, as they demand large amounts of new resources.

Substantial Mineral Mining: As quoted earlier with the prediction of Goldman Sachs, an exponential increase in the demand for metals and other minerals is expected to satisfy the electricity needs of our modern lifestyles and EV is an important contributor to this rising demand.

***

Despite the importance of understanding the overall environmental impacts, I have yet to encounter any data that comprehensively addresses the factors mentioned above. Additionally, the previous contribution of 1.5% to 2% of carbon benefit in power generation does not take into account other complex factors and variables. Further examinations are presented in the following discussions:

Power Generation Principles

Solar's power generation efficiency with modern technology is in the range of 15 ~ 20% and wind offers a higher range between 35 ~ 45%. When the energy sources like sunlight and wind are abundant, the generation efficiency may not be a primary concern. However, the materials used to build solar panels and wind turbines contribute to inefficiency when the opposite weather conditions occur and one of the most challenging factors in these new renewable energy sources is the inconsistency of the weather conditions. Facing such challenges, many complexities including Energy Saving Storage (ESS) systems, backup systems from traditional energy sources, etc., are implicated, further contributing to the inefficiency.

While the generation efficiency of crude oil means straightforward conversion to the refined outcome and does not vary much from one refiner to another, the efficiency of the solar and wind can vary wildly from one energy farm to another based on various factors such as weather conditions, material quantities used to build the farms, energy transportation infrastructure, and many more.

Importance of this perspective is often downplayed by many stakeholders and proponents. However, the total quantity of the materials and resources required for this EV initiative is certainly not "down-playable". It is a truly massive undertaking creating significant impact on the environment.

Fossil Fuel Conversion Efficiency: Fossil fuels, used in ICE, convert chemical energy into mechanical energy to power the vehicles. During this conversion process, a significant amount of energy is lost as heat. Despite advances in the fuel efficiency, which is now typically between 20% and 30%, commonly translated into such units as miles per gallon (MPG) or liters per 100 kilometers (L/100km) for consumers, the laws of thermodynamics limit further improvements in today's engine technology. This means that ICE systems lose 70% to 80% of the energy contained in fossil fuels.

Solar and Wind Conversion Efficiency: Energy generated from solar and wind farms must be transported to urban centers, a process that incurs energy losses roughly estimated to be between 6% and 14%. Additionally, electric motors in EVs experience energy losses when converting electrical energy into mechanical energy, with efficiencies typically ranging from 85% to 95%. Considering these factors, the overall energy efficiency from the solar and wind power generation to the vehicle movement ranges from 73% to 89%.

While the energy efficiency of solar and wind connected with the mobilization of EV appears much higher at face value in comparison with that of ICE, we must consider that it is only for 1.5 ~ 2% of the carbon contribution as of 2022. And there is much more to consider.

Weather Factor: Weather variability also plays a significant role in the EV performance. As the massive conversion to EV is recent, the public is still learning about these effects, as demonstrated by an incident in the winter of 2023, when many EVs were stranded in an unusually cold weather, some on the roads and others in parking, due to accelerated power drainage in cities like Chicago and New York. Extreme weather conditions in both high and low temperatures can create large impacts on driving distance and accelerate self-discharge rates in EVs. While the gas in the fuel tank of an ICE vehicle will remain usable for many years and maintain its fuel efficiency almost irrespective of temperature, EV batteries can experience large fluctuations in mileage and self-discharge rate during the periods of non-use.

Other Important Factors: While the shift towards solar and wind is likely to persist, this trend does come with environmental consequences. Simply assuming that "*the benefits outweigh the negatives*" without a thorough understanding would be misguided as failing to address the associated "negatives" could hinder efforts to maximize the potential benefits in the long run.

For instance, solar farms require substantial amounts of water to cool the systems and clean the solar reflectors, which is critical for maintaining power generation efficiency. The Noor Ouarzazate Solar Complex in Morocco uses an estimated 2 to 3 billion liters of water annually, primarily sourced from the nearby Mansour Eddahbi Dam, located 12 kilometers from the facility. Since solar projects are often located in regions with strong sunlight and long durations, many new installations are in arid and semi-arid climates where water resources are already scarce, as is the case in Morocco.

https://en.wikipedia.org/wiki/Ouarzazate_Solar_Power_Station

Moreover, industrial-scale wind turbines are often criticized for generating excessive noise and posing risks to local ecosystems, affecting both human and wildlife habitats. Similarly, large solar farms can lead to the loss of agricultural land, habitat destruction, and the fragmentation of local ecosystems. They may also contribute to the heat island effect due to sunlight reflection, while shading from solar panels can disrupt natural ecosystems, particularly in marine environments. To address these concerns, a growing trend is to locate solar and wind farms in remote inland areas or away from coastal regions, distant from human populations. However, this shift increases the need for infrastructure to transport energy, leading to higher material demands and greater electricity losses over long distances.

***

The environmental impacts of fossil fuels are well-documented, thanks to centuries of its history, providing an abundance of reliable data across various aspects of the industry. For instance, carbon impacts have been comprehensively analyzed for different sources of crude oil, natural gas, oil sands, shale oil, and more. The straightforward conversion from the crude oil to the vehicle fuel makes these data reliable and transparent.

In contrast, the relatively recent emergence of the EV industry, coupled with its more complex supply chain, makes it challenging to fully grasp its true environmental impacts. With trillions of dollars at stake in the EV transition, there has been a surge of efforts to justify it. These efforts often involve presenting benefits with a limited scope, such as the earlier example claiming a 50% improvement in PM2.5 from EV fleets in the USA. Amidst this flood of selective information, finding truly comprehensive analysis becomes difficult.

As a result, reliable data on EVs' true carbon footprint and overall environmental impact are significantly lacking. Although I have attempted to present some comparative figures in this discussion, my intent was more to address the needs of the readers who may have preconceived notions about the environmental impacts of EVs. In my view, comparing the impacts between the two vehicle types is almost meaningless at this time due to numerous unknowns and unreliable data flooding in the scene of EV. While in-depth discussions will be presented shortly, concerns about the toxicity from the mining activities alone could pose a considerable obstacle to any effort in favor of a widespread EV adoption.

Increasing Power Demand

Artificial Intelligence (AI): As we enter the era of the fourth industrial revolution, marked by the advent of Artificial Intelligence (AI), experts predict that the electricity consumption will rise at a rate significantly outpacing the development of the power generation. The power consumption of servers and data centers required for operating AI systems is exceptionally high. In 2023 alone, it is estimated that the sector required approximately 9 terawatt-hours of energy. This figure is nearly three times the total accumulated power generation capacity of the solar and wind energy sources combined in 2023.

Consequently, several major companies that previously committed to achieving carbon net-zero status within specific timelines have since revoked their commitments due to the escalating power demands posed by AI technologies.

Rapid Increase: The power demand of the regenerative AI emerged in a matter of a couple of years whereas the total power output of wind and solar is the culmination of over two decades of extensive efforts, including various government subsidies, tax benefits and other forms of incentives. As we witness the concurrent evolution of the EV and AI, it is evident that other essential industries, such as construction, healthcare, agriculture, etc., as well as our lifestyles will likely continue to revolutionize in similar directions of exacerbating power shortages.

Supply Challenges: The pace of the solar and wind power generation development is insufficient to meet the growing power demands. This power shortage issue is compounded by geopolitical conflicts that affect energy supply stability. For instance, the energy shortage in Europe following the conflict between Ukraine and Russia has led several countries to restart their coal power plants to mitigate the shortfall. Consequently, the use of fossil fuels is expected to rise, counteracting efforts to reduce carbon emissions and unfortunately, the EV conversion is a significant contributor of such trend.

Careful Planning and Lack Thereof: Given the significant gap between the supply and demand, it is critical to engage in careful planning to allocate available resources while considering the environmental impacts of the associated activities such as mining and toxicity creation. However, such efforts in the global scale are absent. The society is merely going with the flow, with once-closed coal power plants being fired up again while EV production accelerates. In my view, this phenomenon clearly demonstrates that we were not prepared to implement the EV conversion on the scale we are witnessing now, spending hundreds of billions of dollars in the public funds for subsidizing EV industry players and affluent consumers in the Western world.

B. Material Requirements

Higher Demand: While a dedicated section titled **D. Danger of Particulate Matters** will address other related perspectives, this discussion centers on the significantly greater weight and material demands of EVs compared to their ICE counterparts. Estimates indicate that EVs are 10% to 30% heavier than similarly classed ICE vehicles, leading to a proportionally higher demand for various materials. Since each ounce of the material we use leaves an environmental footprint, this considerable difference has significant implications on the environment.

As in the example of the copper mentioned earlier, many other metals and minerals are similarly impacted. These include lithium, cobalt, nickel, manganese, graphite, and rare earth elements (REEs) such as neodymium, dysprosium, and praseodymium, which ICE vehicles require little to none.

The increased weight of EVs is primarily due to the significantly heavier batteries, which include materials in cells and modules, cooling system, wiring, electrical management, and casing systems. The energy density of the current battery technologies is considerably lower than that of gasoline, necessitating more material to store a comparable amount of energy. For example, typical lithium-ion batteries have an energy density of about 150 ~ 250 Wh/kg, whereas gasoline has a much higher energy density, approximately 12,000 ~ 13,000 Wh/kg. This substantial difference of 5 to 10 folds results in the need for much heavier batteries to achieve similar driving ranges.

A particular concern arising from this scenario is the extremely high demand for the materials extracted from the Earth's crust, the toxic nature of these materials, and the environmental impacts associated with mining them.

Elevated Environmental Harm: Batteries are made with highly reactive materials designed to generate electrical energy through the movement of charged ions; anions (negatively charged) and cations (positively charged). As discussed earlier with pesticides, fertilizers, dyes, and other chemicals, increased chemical reactivity typically correlates with greater environmental impact and toxicity.

For instance, rare earth elements (REEs) were once mined in developed countries like the U.S. and Canada, but many of these mines were shut down due to a particularly high level of pollution and toxicity associated with extraction and refining processes, including radiation exposure from radioactive elements to workers and nearby residents. While all mining activities create significant environmental impacts, they are typically tolerated even in these developed countries mostly because of the significant economic gain. But, in case of the REE mining, the impacts were too large and could not be tolerated.

As the REEs are crucial for a wide array of industries, including electronics, renewable energy, medical devices, lighting and more, in addition to EVs, production had to shift elsewhere. China, with its rapidly growing manufacturing sector, filled this gap and now accounts for over 95% of the global REE production. However, recent trade tensions between China and other nations have begun to reshape the landscape of the REE mining, including many ongoing reshoring projects.

C. Toxicity Factor

Many people may lack experience or knowledge regarding the environmental impacts of the mining industry. These impacts are profoundly concerning, yet largely overlooked, despite their global nature.

For instance, the Columbia University recently published an analysis highlighting the environmental damage caused by lithium mining, a crucial component for batteries used in all EVs. The report, titled "*The Paradox of Lithium*," underscores the severe environmental harm associated with the lithium extraction and stated, *"we must find alternative to Lithium."*

https://news.climate.columbia.edu/2023/01/18/the-paradox-of-lithium/

As of this writing, there have been ongoing discussions about emerging battery technologies in the EV industry, such as solid-state and lithium-free, among others. However, it is clear that none of these new battery technologies can considerably reduce the harmful environmental impacts associated with the mining activities as they all rely on the materials extracted from the Earth's crust. Eliminating one mineral will likely require substitution with another, marking a similar net effect.

Mining and Concentration

Mining is, simply put, the process of extracting specific minerals from the Earth's crust. For example, when mining for copper, miners dig into areas where copper is present in low concentrations and remove other materials to isolate and purify the copper to a desired level.

The environmental impacts of the mining have evolved significantly over time. Historically, metals and other desired minerals were abundant on the Earth's surface, enabling early civilizations to create the Bronze and Iron Ages without heavy equipment used in today's mining. They simply collected visible rocks containing high concentrations of metals and smelted to purify them.

Low Yield Percentages: Today, however, mining operates on a vastly different scale due to the depletion of these easily accessible resources. To obtain 1 kg of copper, for instance, modern mining operations may need to process 100 kg to 200 kg of ore, yielding a 0.5% to 1% extraction rate. The rate continually decreases as a result of the ongoing heavy industrial mining all over the world.

Concentration of Toxicity: The primary issue with the modern mining is that concentrating target materials also concentrates other toxic elements present in the Earth's crust, known as "ore" in the mining industry. These elements include mercury, arsenic, lead, cadmium, etc., which are highly reactive and toxic in high concentrations, while some sites create concentrated radioactive waste such as uranium. In their natural, undisturbed state, these elements are typically dispersed in low concentrations within the Earth's crust and pose minimal threats to human health. However, when disturbed and artificially concentrated through mining activities, they become significant environmental hazards.

Toxicity Impacts: Communities near mining sites often suffer from exposure to highly contaminated air, water, and soil, leading to increased rates of infant mortality, reduced life expectancy, poisoning, and cancer among many other related diseases. Surface residues and runoff water continue to contaminate the environment and spread through the circulation of water and air globally.

One notable example is the Berkeley Pit in Butte, Montana, a former open-pit copper mine. After the cessation of its operations in 1982, runoff water accumulated in the pit, creating a toxic lake.

Fig.57 Mining Toxicity

Berkeley Pit (center) and Yankee Doodle Tailings Pond (upper left) with terraced levels/access roadways. The city of Butte is at lower right.

https://en.wikipedia.org/wiki/Berkeley_Pit

In 1996, the discovery of 342 snow geese carcasses on the lake's surface prompted investigations, revealing that the birds had died from exposure to acidic, metalliferous water. Efforts to prevent birds from landing on the contaminated water have been ongoing, yet in 2016, thousands of snow geese died despite the efforts.

Toxins from mining activities persist in the environment for extended periods, often exceeding human lifespans in multiples.

Toxicity Spreading around the World

If one assumes that the toxicity issues from mining are confined to local areas and do not affect urban populations, it would be a large mistake.

Blood Lead Levels (BLL) in the general population are a clear indicator of such harmful chemicals from remote sources impacting humans all around the globe. The following excerpt from the Wikipedia provides more context on this issue:

Blood lead level (BLL), is a measure of the amount of lead in the blood. Lead is a toxic heavy metal and can cause neurological damage, especially among children, at any detectable level. High lead levels cause decreased vitamin D and haemoglobin synthesis as well as anemia, acute central nervous system disorders, and possibly death.

Pre-industrial human BLL measurements are estimated to have been 0.016 µg/dL, and this level increased markedly in the aftermath of the industrial revolution. *At the end of the late 20th century, BLL measurements from remote human populations ranged from 0.8 to 3.2 µg/dL. Children in populations adjacent to* *industrial centers in developing countries often have average BLL measurements above 25 µg/dL.*

https://en.m.wikipedia.org/wiki/Blood_lead_level

According to the information, the astonishing 50 to 200 times increase for remote human populations was the case from the pre-industrial era by the end of the 20th century. For children living near industrial centers, the increase is even more alarming, exceeding 1,500 times higher. The Wikipedia article also quotes any detectable level can cause neurological damage, especially for children.

Similarly with the lead, many other heavy elements and compounds are impacting our lives – more in-depth examination will be presented in **Government Work** discussion later in this subsection.

Lack of Human Ability: When discussing the BLL in particular, it is crucial to acknowledge a pivotal moment that resulted from a poor decision made by the petroleum industry decades earlier. This decision significantly accelerated the rate of the lead bioaccumulation in humans and animals. Although later rectified, the damage had already been done - More details on this incident are provided in the **Clean Recycling Initiative™** section later in this chapter. Unfortunately, the rectification has not brought down the BLL to a level anywhere close to the pre-industrial era and it is still tens or hundreds of times higher for people all around the world.

I highlight this incident as it serves as a clear example, among numerous others, of the environmental harm created from poor human decisions. It is important to recognize that these decisions were often made without fully understanding their detrimental effects on humanity and it is sadly the nature of humans lacking ability.

Time Delays and Prolonged Impacts: Another example, similar to the case of the BLL, is the use of asbestos. The harmful material was used for many decades in constructing homes and buildings before it was stopped in many countries as the severe health damages were finally understood. Unfortunately, however, it still remains in older buildings in high concentrations, continually releasing into the environment, spreading via air and water and contaminating and accumulating in the living organisms including humans and animals.

Conversely, there are many other known examples which highlight human's inability to understand the magnitude of the impacts of certain decisions and actions until they become abundantly clear, resulting in countless deaths and illnesses of the past, present and future. Even today, it is easily plausible or even certain that many similar incidents continually occur without our knowledge.

Government Work: Health impact from the accumulation of toxic substances in human body is complex and often subject to debate. Various governments and health organizations worldwide provide guidelines on the allowable intakes of some notable toxins, expressed in different units of measurement.

For instance, the World Health Organization (WHO) suggests that a tolerable weekly intake of 1.6 micrograms of mercury per kilogram of body weight. In the meantime, the USA Environmental Protection Agency (EPA) sets a reference dose (RfD) of 0.1 micrograms of the same metal per kilogram of body weight per day, which calculates to 0.7 microgram per week, less than half of the WHO's tolerable weekly intake.

General Interpretation: Upon facing such information with a large discrepancy between the two highly authoritative organizations in the subject, I find it extremely challenging to process it.

With my background as engineer and many years of professional experiences and personal observations, the prevalence of harmful chemicals in the environment including food, air, water and everything around us is clear to me. However, I realize that this concept may not be as intuitive for many others in the general public. The fact that influential organizations provide these guidelines serves at least an educational purpose, highlighting the presence and risks of these substances.

Interpreting these guidelines and taking necessary actions, however, such as identifying the sources of exposure, measuring and controlling the intake, and attempting to avoid them, would be an impossible task. It is akin to trying to avoid contact with other invisible harm such as viruses and bacteria, unless we live in a bubble without ever eating.

Furthermore, what do "allowable" and "tolerable" mean? Science tells us that many harmful substances accumulate in the body and remain there for a lifetime. If we consider the BLL of the pre-industrial era as the baseline for the natural presence of these elements in our bodies, how can we justify an "allowable" or "tolerable" deviation from this baseline, especially when our bodies already contain levels that are tens or even hundreds of times higher than the baseline?

Most often, such guidelines with "allowable" or "tolerable" limits are associated with certain conditions, for example, applicable to "healthy" children or adult, who do not have diseases or infections which can create compounded impacts. However, we must recognize that the term "healthy", or any other words or expressions utilized to validate such claims, is such a vague term that can involve numerously different personal lifestyles and health conditions.

My Own Interpretation: Given my understanding of the toxicity levels of various harmful substances, their persistent bioaccumulation in the body, and the complexity and individuality of each person, I find the notion of the "allowable" or "tolerable" limits misleading at best. Furthermore, I am skeptical that even influential and powerful organizations such as the WHO and the US EPA fully grasp the compounding effects of various toxic elements and compounds.

This skepticism stems from the following two key factors either proven or highly plausible:

1 **Personal Variability:** The health impact of a specific cause can vary significantly between individuals (even for "healthy" ones).

2 **Authoritative Errors:** Governments and authorities often make significant errors. As reviewed earlier, the WHO's intake limit for mercury is more than two times higher than that of the EPA. Such stark difference between the two leading authorities suggests that at

least one of the limits may be erroneous while, in my view, both are.

These toxic substances are present in food, drinks, and even air. There are hundreds or even thousands of toxic elements and compounds around us. They are not visible contaminants that can simply be removed from our consumption. Avoiding these substances is impractical unless we cease living.

From a scientific and analytical perspective, it is implausible that authorities can fully understand the cumulative effects of all these substances on different individuals. Consequently, I disagree with the practice of posting such vague limits, as it creates a false sense of security. This practice is, in my view, irresponsible at best.

D. Danger of Particulate Matters

The issue of toxicity extends beyond the mining activities and includes concerns over the increased quantities of the particulate matter (PM) emissions from EVs compared to their ICE counterparts. The nature of this issue is both profound and serious.

Weight Difference: Due to the heavy weight of the batteries, EVs are generally heavier than similar classes of ICE vehicles. Estimates of this weight difference vary, with some suggesting an increase of 10 ~ 15%, while others propose up to 30%. Although it is challenging to judge the accuracy of these estimates as vehicle manufacturers do not typically make two directly comparable vehicle models, it is evident from the related sciences and reliable industry sources that EVs are indeed considerably heavier than their ICE equivalents.

Particulate Matters of Vehicles: A blog published in the Yale University's Environment 360 website discusses the increased quantities of the particulate matters in the air from EV's tire friction against road surfaces, attributed by the higher weights. It is indeed a basic principle of the physics that greater mass generates stronger forces, leading to a higher likelihood of things breaking. This blog highlights the toxicity of a specific chemical known as 6PPD. According to the article, "*Tire emissions from electric vehicles are 20 percent higher than those from fossil-fuel vehicles*"

https://e360.yale.edu/features/tire-pollution-toxic-chemicals

The blog also mentions that "*Seventy-eight percent of ocean microplastics are synthetic tire rubber, according to one estimate*," a highly alarming figure.

Health Impact: The dangers of the particulate matters are well-documented. While the Yale Environment 360 article addresses the toxicity of 6PPD, it is also known that even the non-toxic particulate matters such as invisibly small rubber particles can accumulate in body organs, tissues, and even brain, causing adverse health effects such as heart attacks from the blockage of blood vessels, cancers, etc. This is not a "down-playable" factor in our lives. It seems paradoxical that the global focus is almost exclusively on reducing carbon emissions from the vehicles' exhaust pipes while neglecting the potentially more harmful and immediate effects of particulate matters from the increased weights of EVs.

***

Interestingly, the Yale blog starkly contrasts a popular belief that plastics, particularly synthetic fibers, are the main sources of microplastics, by reporting 75% of the microplastics in the oceans are from vehicle pollutants. Personally, I have a difficulty to comprehend how it would be feasible to calculate the composition of the particulate matters, as discussed earlier. In my analysis, microplastics are so small that even the highest quality of filters cannot catch them all. Although I have not had a need for analyzing microplastics in the oceans, it would be natural to assume that heavier and larger rubber particles are captured easier than lighter and smaller dust from other plastic sources. Perhaps, it would merit to review the full scope of the study utilized to conclude such high percentage of 75%, but unfortunately, it was not made available.

Nevertheless, I quote this information as we can safely conclude one common phenomenon: Increased weights of EVs lead to more pollutants of particulate matters in our environment and they are highly impactful for human health.

E. Water Consumption and Contamination

Mining and refining processes consume vast amounts of water. For example, copper mining and refining involves melting ore at high temperatures and using large quantities of water and chemicals to separate the target metal through flotation. During this process, water is agitated to create bubbles that carry copper particles to the surface, leaving other materials behind.

As established earlier, this process not only requires a substantial amount of water but also leads to significant contamination with toxic elements such as lead, arsenic, mercury, and cadmium. The contaminated water, known as acid mine drainage, spreads around the global ecosystem and poses severe environmental risks worldwide.

F. Questionable Recycling

Two principal methods of battery recycling exist; destructive and non-destructive. The non-destructive recycling involves careful examination of the residual usefulness of the components and chemicals remaining in used batteries without breaking them down completely. This concept is akin to the refurbishment of electronics or segregating useful parts from the waste and is more desirable form of recycling as it leaves minimal environmental impact.

More problematic method is with the destructive recycling. As in the case of the conventional plastic recycling, it involves the crushing and grinding of waste batteries on the outset of the recycling. Drawing from my investigative work on the plastic recycling methods, as reviewed in **Chapter I: Unfortunate Recycling**, it is evident that the application of heavy physical forces to crush these materials generates substantial quantities of dust.

Dust Compositions: In the context of the plastics, there are important advantages compared to the batteries. Plastics processed in recycling facilities are primarily composed of similar and chemically inert materials. While plastic dust can accumulate in the working environment and be inhaled by the workers, its inert nature poses fewer immediate health concerns. In contrast, the dust produced from crushing batteries, containing a variety of chemically reactive substances, including metals, REEs, etc., poses much higher level of environmental and health threat. Considering that batteries are designed to move electrically charged ions to create electrical energy as mentioned earlier, one can easily understand highly reactive nature of the chemicals contained in them.

These substances can be toxic when inhaled and accumulated in the body, potentially leading to detrimental effects in a near to medium term. Alarmingly, the processes I witnessed did not contain the crushing and grinding process in enclosed environments, raising concerns about the health and safety of the workers involved, not to mention serious contamination far and wide from invisible reactive dust in nearby and even remote residents.

Recycled Battery Performance: Contaminants in the recycled plastics lead to considerably lower strength and overall performance in the recycled fibers, creating and shedding more microplastics and chemicals during manufacturing and consumption compared with the virgin fibers. Similarly, the destructive method observed in some battery recycling facilities, which involves various components with widely different characteristics, suggests that the degree of contamination in the recycled materials could be much more significant than that of the plastics. As in the case of the recycled fibers, assuming that recycling batteries is sustainable with no hidden side effect and downplaying potential harm based on the simple word description, "recycled", would be a large mistake. Additionally, recycled batteries containing contaminants may exhibit significantly reduced performance, potentially undermining the benefits.

G. Unknowns and Certainties

Although the invention of the electric vehicles dates back to the early 19th century, the massive EV conversion we witness today is a relatively recent development. The global EV sales reached the milestone of 1 million units only in 2017, accounting for a mere 1% of the total car sales worldwide at the time. By 2024, the EV sales are expected to constitute approximately 14% ~15% of the global car sales, reaching 17 million units. In terms of the total number of the vehicles on the road, EVs currently represent roughly 3% according to the International Energy Agency (IEA).

Given the rapid increase in the new vehicle sales and substantial financial investments involved, the EV stakeholders must find every means to justify the continuing conversion efforts.

Historically, accurate information regarding the total impacts of a new industry, particularly those as transformative as the EV, only becomes available once the industry reaches a sufficient level of maturity. For the EV, this maturation process may take many more years, during which the broader public may only gradually become aware of various negative impacts. While there are too many unknowns and serious concerns unanswered in this massive movement, we certainly know that it is virtually impossible to remove toxins from the environment once created. With this point of view, I present my proposal of conducting retroactive analysis work, explored in **21.3 Retroactive Analysis** discussion later in this chapter.

20.2.2 Cost and Effects

The amount of money and effort our societies have dedicated to the EV conversion is unprecedented. In fact, this communal endeavor may be one of the most significant financial undertakings in human history especially when considering a very specific objective of the carbon impact it pertains.

Scope of Cost Structure: The scope of this investment extends far beyond the cost of manufacturing the EVs themselves. It encompasses the entire supply chain, starting from the mining of metals and minerals, the manufacturing of solar panels and wind turbines, the installation of energy farms, the grid infrastructure construction and charging stations, the construction of manufacturing facilities for vehicles and batteries, and countless more. The total financial commitment to the EV conversion, when accounting for all related expenditures, likely amounts to tens of trillions of dollars.

Effectiveness of Efforts and Support: Given that technological advancements typically correlate with the level of investment, there is a high probability that more effective solutions could be developed that do not produce such harmful environmental side effects in my view. Many technologies have been proposed and tested with limited resources and support. For instance, the carbon capture technology has proven largely ineffective due to the low density of carbon at emission sources as well as in the air away from the sources. However, with sufficient resources and investments, these challenges could potentially be overcome, yielding more cost-effective and time-efficient solutions. This technology would also be valuable to capture methane (CH_4) release from thawing permafrost, a topic that will be explored further in the following discussions.

More immediate solutions, like the thermally insulated curtain discussed in **Chapter I: 2.6 Other Important Textile Applications and Environmental Impacts**, could potentially achieve a more significant reduction in the carbon emissions than the estimated 1.5% to 2% carbon savings from transitioning to EVs in the power generation, and at a much lower cost. Moreover, implementing this solution could reach a significantly larger number of consumers in a fraction of time compared to EVs. Depending on the scale of deployment, it could take just months to reach as many households as there are EV owners worldwide, whereas deploying EVs and building necessary infrastructure for solar and wind energy is a much slower, more complex and environmentally harmful process.

Unfortunately, many promising ideas with incomparably smaller environmental footprint, fail to attract sufficient attention or support due to the narrow and limited visions of the global decision makers, combined with the efficient efforts made by the stakeholders and proponents of EV to divert the public's attention from more effective efforts.

Subsidies from Public Money: Governments around the world have provided substantial subsidies, tax credits and other forms of incentives to both EV consumers and other stakeholders in the EV industry. For instance, in 2021 alone, $30 billion was allocated globally to subsidize EV purchases in affluent countries for affluent consumers. Furthermore, hundreds of billions more have been provided to industry players in various forms of incentives. Given that these funds are sourced from public taxation, it is crucial to evaluate the overall benefits and drawbacks of such extensive spending. This includes whether alternative investments might offer comparable or superior returns for the public good. In my view, all incentives must stop until we have a clear understanding.

EV Discussion Conclusion

This subsection underscores numerous factors that must be considered when evaluating whether our global society is heading in the right direction regarding the environmental management and sustainability, using the EV conversion as an example.

Humanitarian Impact: Any effort toward sustainability must also account for humanitarian perspectives. This consideration may seem mundane to many, but the humanitarian crisis in many countries where EV-related economic activities occur reveals its critical importance. In some regions, minerals are extracted through "artisanal mining", a term that belies the horrendous and dangerous conditions involved. Workers, including children, excavate targeted minerals with bare hands without any personal protective equipment, not even simple face masks. They face significant health risks and live in inhumanely poor conditions.

Despite billions of dollars circulating among EV stakeholders, including affluent consumers receiving government subsidies for their EV purchases, these artisanal miners are not paid enough to afford even the most basic safety equipment. This stark disparity raises critical questions about the sustainability of the EV conversion. If the conversion is truly beneficial for the humanity, then the miners should be provided with proper personal protective equipment and fair wages.

Many argue that the benefits of the EV conversion outweigh the negative impacts. However, this assertion merits serious scrutiny. In my view, the toxicity aspect alone could easily outweigh the benefits of the conversion. Unlike such simple assertion by the proponents, I remain unconvinced that there is a net benefit to the EV conversion, therefore my vehicle choices will remain as ICE.

Section 21   Solutions

Throughout this book, we have explored a wide range of topics related to textiles and their environmental impacts, starting with the consumer focus in Chapter I. Chapter II provided a deeper understanding of the underlying science, followed by Chapter III, which conducted the Lifecycle Assessment across the 16 evaluation criteria. Chapter VI then tackled the issue of widespread misinformation, identifying it as a major barrier to more effective global environmental management. Earlier in this chapter, we expanded our discussion beyond the textiles by examining the complexities of the EV conversion, further emphasizing the need for a more holistic approach to addressing the significant environmental challenges we face.

From these discussions, it should now be clear to readers that overly simplistic notions like "*natural fibers are more sustainable than synthetic fibers*" do not lead to positive outcomes. To effectively manage our environment and ensure sustainability, global society must acquire necessary knowledge grounded in valid scientific principles.

21.1   Education, Education and Education

The developments of our societies are determined by the level of education of the constituents. Numerous textile companies engage in greenwashing because the public is not sufficiently educated to recognize and challenge these practices. Surprisingly to many readers, even industry professionals with decades of experiences often lack necessary knowledge due to the systematic changes in the textile industry's structure, as discussed in **Chapter IV: Organisations Lacking Expertise**. Often, those spearheading greenwashing activities simply do not know better, underscoring the importance of proper education.

In this discussion, I propose educational programs which will enhance the aptitudes of the society as a whole as well as the individual societal members to be able to make informed decisions for humanity's future. Three critical areas for maximum effects and the inclusion of as many members of our society as possible are addressed **At School, At Home** and **At Government**:

At School

Given the critical nature, relevant curricula must be provided at the school level, particularly at the stages where the benefits can extend to the largest population of the students of the world. This will ensure that future generations make more scientifically sound decisions for the environment.

Although I have never been an educator or developed an educational curriculum within a school system, my lifetime learning experiences in the subject matters inform my views on potentially beneficial programs. Given a proper context and necessary details, I am confident that skilled educators can develop effective programs tailored to their students' knowledge levels and learning abilities. Here are some key subjects I consider important:

- Recycling principles of different materials such as plastics, textiles, metals, glass, etc., and actions required to maximize the purity of the recyclable materials

- Analytical mindset in relation with various sustainability claims and to recognize potential greenwashing
- Nature of toxicity creation and spread around the world
- Other relevant subjects

While students on different learning curves may not yet have necessary analytical skills required to fully evaluate numerous sustainability claims they face, fostering an inquisitive mindset is crucial. Encouraging students to question and study these claims in their current abilities can help them develop their own analytical and self-educational skills as they progress in life. Additionally, hands-on experiences, such as site-visits, simple experiments, or even viewing educational video footages, can effectively teach young minds about the basic recycling principles. Therefore, educational programs designed to cultivate these skills do not need to be highly technical or complex, nor will they require a high level of time commitment or financial investment.

Special Mention – International Sustainability Marketing Competition (ISMC): Recognizing the lack of essential education in the environmental sustainability concerning textiles, I have been offering various educational opportunities for both students worldwide and professionals in the industry. For the industry audience, I have conducted many rounds of webinars on critical subjects and will continue to do so. For the students, the two organizations I manage, **Clean Recycling Initiative™** (Non-Profit) and HEAT-MX™, collaborate to organize the annual International Sustainability Marketing Competition (ISMC).

Despite the term "Marketing" in its name, the ISMC invites students from college level to doctoral candidates, irrespective of their fields of study, based on the premise that everyone consumes and generates significant quantities of textile waste throughout their lifetimes. In the third iteration of the event currently being organized as of this writing, more than 500 schools from 110 countries are participating, highlighting the keen interests of the students around the world, regardless of their regions, backgrounds and lifestyles, on the subject of environmental sustainability.

As a mandate of the event, the participants must attend at least one of the total 7 information sessions, offered in three major time zones: North and South America, Europe and Africa, and Asia-Pacific regions. These educational sessions provide the same content, focusing on two major segments: 1. Examples of greenwashing practices and analytical skills to recognize them, and 2. Recycling principles of textile goods along with solutions offered by the non-profit **Clean Recycling Initiative™**. Then, the participating students are asked to create an effective public campaign to share their knowledge. More details can be found in the company's website below.

https://cleanrecyclinginitiative.com/

Special Mention – Textile Programs in College Level and higher: In **Chapter I**, I discussed several examples of undesirable textile products and technologies, including Artificially Conceptual Values (ACV) lacking real-world benefits and impractical performance metrics habitually adopted for unwarranted consumer goods. It is particularly concerning due to the direct environmental footprints these goods create during manufacturing, consumption and disposal.

Unfortunately, many professionals who design and lead these efforts often lack comprehensive knowledge in related scientific fields. For instance, thermal technologies are seldom included in the curricula of most textile-related education programs, despite their significance in both protective nature and environmental impacts related to the elevated quantities of materials as well as multiple technologies utilized. My search around the globe for educational programs that adequately cover these topics has yielded no results, whether in engineering or fashion disciplines.

Consequently, many professionals in the industry rely on established brand recognitions on certain thermal insulation materials and technologies or habitually follow decisions made by their predecessors or related trends in the industry. This practice leads to judgments and actions based on notions rather than proper scientific understanding. Given the considerations, I propose the following subjects to be included in the textile programs:

- Science of textile materials in relation with the principles of recyclability
- Scientific understanding of performance / maintenance features and their impacts on the environment

- Application-specific technologies such as clothing, footwear, glove, home textile, outdoor gears, etc., and the avoidance of using unnecessary materials and technologies

- Greenwashing definitions and avoidance

- Other relevant subjects

Given that this area overlaps significantly with my educational and professional background, I have developed an educational program with 35 separate subjects that cover a wide range of topics, from basic principles of textile materials to manufacturing, maintenance, disposal, and environmental impacts while continually expanding this list with additional subjects. If any educator in the field of textile reading this book is interested, I would be delighted to discuss further and help.

Special Mention – Sustainability Programs in Post-Secondary Educations: I have recently had an opportunity to work with two individuals, one of whom is a recent graduate of an undergraduate sustainability program at an internationally reputable university and another studying in a graduate program at a different, but equally reputable institution. Through several initial conversations with them, I realized a significant lack of scientific education in these programs. At the level of science related to many topics discussed in this book, it became apparent that neither student had opportunity to develop their analytical skills, nor were they exposed to even basic, yet crucial scientific concepts related to environmental sustainability.

When these students graduate and assume important positions in the fields of environment and sustainability in government-organizations, -affiliates, NGOs, and private companies, they are likely to be easily influenced by notion-based theories from their superiors or influencers in the field and may make unscientific decisions. Given the direct impact these positions can create on global environmental efforts, it is critical that higher education programs specialized in these environmental sustainability subjects include curricula designed to develop students' scientific aptitudes and analytical skills.

At Home

At home, the educational levels of household members must be sufficient to understand recycling actions on consumed goods necessary to support positive environmental outcomes. Only through collective societal efforts on post-consumption can we reclaim resources with minimally negative impacts.

To achieve this, households must gain practical knowledge. While basic principles of recycling for plastics and textiles will be presented subsequently, I propose that municipal and federal governments collaborate to provide the households with the following capabilities:

- **Adequate Training**: Households should be sufficiently trained to understand the principles of recycling and how to sort materials to maintain desired purity levels in subsequent recycling processes.

- **Informed Households**: Households should be well-informed about the efficiency of recycling efforts made by local and federal governments, enabling them to collectively engage in more effective environmental efforts.

- **Educational Resources**: Educational resources should be readily available to all household members, regardless of age or educational background, to ensure comprehensive understanding and participation in desired recycling efforts.

Given that households form the foundation of our society's complex fabric and that the effects of proper education can extend to various community and industrial activities, it is crucial to implement effective educational programs at home.

At Government

At the government level, developing the highest degree of expertise is crucial, though not necessarily internally, for guiding citizens toward effective environmental practices. In line with the proposals presented in the next subsection, **Nations Unified for Environment and Sustainability (NUES)**, necessary expertise and responsibilities can be developed and tackled without excessive and time-consuming investments undertaken independently. With this scope in mind, governments worldwide must:

Comprehensive Educational Contents: Supply necessary educational materials for the likes of the tasks identified earlier in "**At School**" and "**At Home**", ensuring that citizens are well-informed about the environmental effects of their actions and leading to more beneficial and impactful participations.

Effective Collection and Processing Systems: Implement a robust system for collecting recyclable materials and processing them, ensuring efficiency and effectiveness in resource reclamation. More details on the processing based on the principles of recycling and its economic models will be discussed in **21.5 Recycling Principles** later in this chapter.

Dissemination of Relevant Information: Share information about governmental and societal efforts for the environmental management, both locally and globally. This includes the bases of the decisions and actions undertaken and impact analysis results on both positive and negative outcomes as well as potentially harmful impacts.

Continuing Education Programs: Establish programs for continuing education, targeting citizens who have not previously had the opportunity to gain essential knowledge on key environmental subjects such as principles of recycling, household sorting efforts in relation with collection systems in place, etc.

Establishment and Maintenance of Educational Sites: Create and maintain educational sites where citizens can gain hands-on experience and knowledge on the principles of recycling, positive impacts created from their sorting efforts and other relevant environmental topics. Simple demonstration equipment or educational video footages may be provided for effective learning – This is one of the winning ideas of the International Sustainability Marketing Competition (ISMC) 2023. Any municipal or federal governments interested in this idea can contact the **Clean Recycling Initiative™** organization for more information.

Demand for Governmental Transparency and Leadership: Be transparent! The nature of political systems often leads governments to highlight successes and downplay failures. Significant social and financial resources are spent on promoting positive outcomes through advertisements, funded studies, etc. We generally accept it as a social component of governance. However, in the perspective of environmental management, it is unrealistic to expect actions that only yield positive outcomes without any adverse effects.

For instance, the EV conversion discussed earlier illustrates the complexity of environmental management, where actions can yield both positive and negative consequences. However, governments around the world do not diligently share the negative sides of the conversion, contradictory to their eager promotion of the positivity and spending enormous amount of public funds.

Strong leadership based on full transparency is crucial; actions must be evaluated in comprehensive scopes, and the public must be educated to understand the overall impacts. Proper education enables citizens to support well-informed government decisions and oppose those that do not align with truly beneficial environmental and sustainability goals. We expect such leadership from the authorities.

21.2 Retroactive Work

As global communities have committed significant resources to various environmental initiatives, often without proper scientific foundations, it is crucial to evaluate their effects comprehensively, albeit belated, and take corrective actions if necessary.

Thorough Analysis on Environmental Impacts

Shotgun Approach: As exemplified by the EV conversion, many powerful organizations spend considerable resources and take actions without adequate analysis. This phenomenon is often motivated by various types of benefits they can gain politically, commercially or operationally. Furthermore, these actions are often supported by the public as a deep level of scientific knowledge to understand the true implications beyond the claims and actions made by these stakeholders is lacking. Unfortunately, this "shotgun" approach is a common occurrence in our society. Most of them are in part supported by public's sentiment of urgency to act.

Despite the broad public support of these actions, therefore, it is imperative to follow through with thorough analysis to ensure that the actions lead us in the right direction. In the case of the EV conversion, such efforts have not been made yet, even though many proven environmental side effects and humanitarian crises exist, not to mention tens of trillions of dollars already spent and myriad more to be spent in the future.

Support Principle: From my perspective, monetary support, such as government subsidies, tax credits and other forms of incentives should only be provided after clear environmental benefits are demonstrated by comprehensive and thorough analysis, encompassing the factors addressed in this book and more. Otherwise, these initiatives should be treated as regular business ventures. Based on the earlier analysis, it is pertinent to question why governments worldwide are providing substantial public funds to affluent EV consumers and business entities in the Western world without confirmed net benefits for environmental sustainability from the conversion.

Retroactive Analysis: Many other environmental initiatives have followed similar "shotgun" approaches, costing global citizens significant public funds. While the public may accept these costs if they result in positive environmental impacts, the history has shown that human actions, with limited scopes and inherent limitations in abilities, often lead to significant environmental harm which can last decades or even centuries. Examples are numerous: The use of Lead in vehicle fuels, The Asbestos in construction, The Perfluorooctanoic acid (PFOA) in non-stick cookware and outdoor apparel, The DDT for pest control, The Chlorofluorocarbons (CFCs) in refrigeration, etc.

Unfortunately, it took decades to understand the negative impacts of these actions on the human health and environment. Lack of thorough analysis, even after implementing such shotgun approaches, can have irreversible effects on our environment and sustainability. It is imperative to conduct comprehensive evaluations to avoid the repetitions of such undesirable human behaviors.

21.3 Responsibilities of Search Engine Operators

Search engine operators have become major sources of information, often replacing traditional textbooks and encyclopedias. The level of trust that people place in the first sets of search results, particularly from platforms like Google, is exceptionally high, often undeservedly so. Over more than two decades, these platforms have consistently satisfied people's information needs conveniently and quickly. Similarly, the recent advent of various AI tools with the ChatGPT at the forefront has gained rapid trust from the public. However, the easy access has created large gaps between the popularity and the accuracy of information on the internet.

Popularity of Misinformation: Due to the prolific activities of self-proclaimed experts in environmental and sustainability topics, the internet is flooded with misinformation. These individuals are often highly skilled in spreading their messages, leading to widespread misconceptions about what is beneficial or harmful to the environment.

Roles of Artificial Intelligence (AI): The revolution of AI has made the society even more reliant on easy access to and popularity of information rather than its accuracy. As of this writing, some experts say that all available data on the internet has already been or will soon be used up in training AI models and machine learning will be more reliant on rapidly growing. As a result, "synthetic data", a term to refer to artificially generated data based on available real-world data or even synthetic data regenerating themselves, will be prevalent on the internet.

As I lack expertise in this area, I am unsure if AI will become sophisticated enough to differentiate between the popularity and the accuracy on the internet. In my observation so far, it may not be the case, at least in a foreseeable future.

My Encounter with Google: In Mach 2023, I was invited as the closing keynote speaker for the Customer Experience Summit in Vancouver, Canada, organized by the Strategy Institute. There, I presented some information discussed in this book. I spoke specifically about the issues of widespread misinformation and its hinderance in global efforts for our sustainability.

Before my session, I was attending other sessions and one of them had a Google representative as panel. Anyone can easily understand that Google would have a lot of insights for customer experiences as it may enjoy the largest customer bases of all businesses around the globe. While I appreciated the opportunity to listen in the expert views on a particular subject the panel was discussing, I could not help but pose a question as it had been lingering in my mind for a long time and was related to the topic of the session.

Question: The question was along the lines of *"People have a strong tendency to trust the information displayed on the 1st or 2nd pages of Google from their search as accurate. But in reality, particularly in the subject of sustainability, widespread misinformation often dominates the internet including Google's search results. Does Google recognize it and have any plan to rectify it?"*

Responsibility: Whether or not Google acknowledges, I believe it must commit to a certain degree of accuracy in the information it displays. While I am unsure if Google would admit their role, many individuals who are prolific in generating misinformation are educated by the search results on its site. Therefore, in my view, Google's responsibility encompasses the educational aspects of the global citizens.

Political Correctness and Missed Opportunity: I recognized that my question might have seemed unexpected and challenging to address. However, the representative responded politely in what I perceived to be a politically correct manner. While her response did not fully address my concerns, the session was shared with another panelist and limited to 30 minutes, preventing further engagement on my part. I then invited her to my session, where I promised to present more comprehensive views of such phenomenon with specific examples and several proposals I had for the company. She graciously accepted my invitation and expressed interest in further discussions during my session. Unfortunately, she was unable to attend, likely due to other pressing commitments.

Belated Proposals: Although delayed, I now present the following suggestions, applicable also to other search engines.

While some search engine operators may claim that they already implement strategies for enhancing the accuracy by providing various tools such as Google Scholar, Google Patents, or ChatGPT's inclusion of information sources in more academic searches, these tools have failed to effectively curb the spread of misinformation.

The issue becomes clear when we consider that many users are unaware of these specialized search tools like Google Scholar or Google Patents. These platforms are typically used by individuals with specific academic interests, while the general public relies on mainstream search engines, where they are often exposed to significant amounts of misinformation. In the case of ChatGPT, simply listing sources of information neither guarantees the accuracy of the content nor ensures that users will delve deeper to gain a more thorough understanding of the subject.

Given that widespread misinformation on these platforms is primarily driven by people's inclination to seek quick and easy answers, search engine operators must adapt their strategies to prioritize and highlight accurate information with this tendency of their users in mind.

For instance, if someone searches for information in their main search tool used by the general public about the sustainability of natural versus synthetic fibers, the search results should provide a comprehensive overview, similar to the 16-point evaluation criteria used in the Lifecycle Assessment discussed in **Chapter III**.

***

I understand that the related tasks of my proposals would be daunting even for large organizations like Google or ChatGPT. It will require these organizations to develop a high level of internal humane expertise in critical subject matters to engineer the platforms that reflect the necessary level of accuracy instead of relying on their existing mechanisms driven by the popularity of information. While I refrain myself from delving deeper, the vector search method and RAG (Retrieval-Augmented Generation) technic, foundational in today's search engine mechanisms, seem to present continuing challenges in information accuracy. Given the importance of the environment and sustainability, such human intervention efforts are warranted in my view.

Furthermore, all involved parties, including the public, search engine operators, government organizations and all other stakeholders of the environment and sustainability, must understand that even the best efforts do not necessarily guarantee the revelation of the absolute truth or only good outcomes. In many cases, long-established theories are proven untruthful with the emergence of new scientific evidence. In other cases, unknown side-effects may become significant with a time delay. Making the best efforts is often the most we can do within our capabilities and is the responsibility of such influential entities impacting our lives. Lacking efforts and relying on the public for their own interpretations of critical environmental information or misguiding them with the misinformation they display is clearly not an example of "making best efforts".

21.4 Nations Unified for Environment and Sustainability (NUES)

Alerted by numerous wrongful claims, actions, and inefficient efforts in the realm of the environmental sustainability, it is my firm belief that a new organization, akin to "**Nations Unified for Environment and Sustainability (NUES)**", albeit under a different name, independent of any political and commercial influences, is necessary. This proposal considers the catastrophic failure from the existing systems including the UN, governing parties and political systems around the world, NGOs, and profit-seeking entities, often operating independently with varying agenda of their own and ineffectively for globally beneficial effects.

Dominance of Inexperience

A significant portion of such failure arises from a substantial disconnect between the perceived environmental benefits held by key decision-makers, often from affluent countries, and the actual real-world impacts. Globally influential decisions are frequently based on theoretical assumptions that fail to account for the realities faced by those most affected.

Gap between Theory and Real-world: As discussed in **Chapter I: "Fast Fashion"**, the European Commission's campaign, "*It is time to put fast-fashion out of fashion*" is a classic example of actions lacking real-world implications. Another recent campaign to ban the destruction of unsold textiles within the EU territories appears also ill-conceived, as it may simply result in more shipments of textile waste to less fortunate countries, unless they can make the unwanted goods magically disappear from the Earth, or even more magically, the campaign organizers can make these initially unwanted textiles wanted again by the consumers within their regions. The later seems implausible to me based on the understanding of human nature, the general shopping habits of consumers, and the impossibility of textile companies being able to produce "only wanted and formerly-unwanted-but-newly-wanted goods".

Thus, the destined outcome of this program may simply be the increased amount of waste shipped to other countries. This is another example of decision-makers from affluent countries lacking the real-life implications of their actions and not fully grasping the realities of life in the neighborhoods of these underprivileged countries where textile waste is pushed to riverbanks or burned in school playgrounds.

Victims' Voices: In reality, affluent countries already ship enormous quantities of their waste, including textiles, plastics, and other industrial waste, to other countries across international oceans just because it is less costly to do so than destroying them locally with strict environmental controls. Without sufficient input from the affected communities in Africa and South America, based on their real-world crises from the waste, such initiatives are destined to fail in creating truly impactful actions. In fact, this principle underpins the decision to maintain 50% of the judges and ambassadors for the annual International Sustainability Marketing Competition (ISMC), mentioned earlier, from underdeveloped and developing countries. This approach ensures to promoting realistic and impactful environmental solutions that benefit all citizens of the world, not just those in affluent countries.

Tools and Capabilities

The proposed "Nations Unified for Environment and Sustainability (NUES)" must include all necessary subject matter experts and scientists who can address pressing issues with proper scientific rigor, real-world impact assessments and actions. This organization should be led by true experts, encompassing the truly comprehensive scope of environmental challenges, including the value chains from farming, manufacturing, consumption, and disposal. In this organization, decision-making must be guided solely by the goal of benefiting the environment, humanity and eco-systems. To accomplish these goals, the organization must be able to:

Independence: Ensure complete autonomy from political, operational, and commercial influences. It is essential that each participating country contributes an equal amount of funding to prevent larger donors from exerting disproportionate influence or expecting greater returns.

Expertise: Work with true experts across all related value chains, whose knowledge collectively extends to interdisciplinary scopes to encompass the entire realm of environmental sustainability.

Global Authority: Manage environmental affairs on the global scale and exercise its authority to lead the worldwide environmental efforts.

Engagement: Collaborate with federal, local, and international organizations for effectively implementing related tasks such as policy implementation, educational program development, transfer of expertise and other related tasks.

***

Implementing these proposals will be undeniably challenging. However, I believe proper education and systematic approach with the clear objectives mentioned above will prevail and lead us to overcome these challenges. For instance, if affluent citizens understand the direct impact of actions like the textile goods they deposit in community recycling bins or donate to charity groups, being ended up somewhere in Africa or South America and burnt with no environmental regulations or monitoring, they will reject these nonsensical practices of shipping their waste across oceans. The proposed organization is tasked to lead the global efforts for such needs.

21.5 Recycling Principles

In the material world we live in, it is virtually impossible to ensure that everything is recycled, upcycled, or reused without any environmental impact. Every item we consume leaves an environmental footprint. Clearly, we cannot cease living for the environment.

A more realistic goal is to minimize this environmental footprint by fully utilizing available technologies to our benefit. A piece of good news is that we can definitely do so much better than what we have been doing around the globe as environmental efforts have unfortunately been largely inefficient, unimpactful and even contradictory and harmful.

While reducing consumption is crucial, equal emphasis must be placed on how to reclaim consumed goods with the highest efficiency of reclamation. In light of this, I present some of the most important principles of recycling, applicable to a large portion of the materials we consume. Understanding these principles will enable our societies to adopt necessary measures to enhance the environmental management and sustainability efforts.

21.5.1 Plastic Recycling

Purity, Purity and Purity

Successful recycling of recyclable materials, including plastics, paper, metals, glass, etc., with minimal environmental impact, heavily depends on the purity of each material type. Pure forms of metals, glass, and thermoplastics can be permanently recycled through closed-loop systems. Similarly, uncontaminated paper can be recycled into higher-value products compared to contaminated paper.

Conversely, any form of contamination, whether similar materials with different characteristics or completely foreign substances, introduces impurities in both recycling processes and final recycled materials. As evidenced in **Chapter I: 4.1 Recycling Myth** section, such impurities degrade the quality of the recycled materials and their future recyclability, while creating many other negative impacts during the entire lifecycles of such materials.

Governmental Efforts and Citizens' Supports

Recycling systems in many countries often have considerable loopholes that affect the sorting and processing of recyclable materials. Comparing the following two different systems effectively illustrates the picture clearly:

Montreal, Canada: In Montreal, municipal governments provide one large bin for the locals in a household or a certain residential proximity to place all recyclable items together into the bin, then transport all the items to a sorting facility. This single-stream recycling approach means that people deposit a wide variety of recyclable materials, such as paper, plastics, glass and metals, into the same bin. Observations of these bins often reveal non-recyclable items such as regular household waste, unrecyclable thermoset plastics, electronics, batteries, light bulbs, etc., all mixed together with other recyclable materials.

Such heavily mixed collections lead to inefficient sorting and the inclusion of foreign materials in the recycling stream. Given that the purity of materials is crucial for successful recycling with minimal side effects, Montreal's recycling efforts are compromised from the outset.

Seoul, South Korea: In contrast, South Korea has implemented a focused approach to public education on waste separation in schools and communities. Following the extensive training, citizens place recyclable materials into designated bins for each material type. They wash off foreign materials such as food residue in containers, liquid remaining in bottles, etc., before placing the materials into the sperate bins.

While the entire collection system impressed me, one particular effort stood out as, in my view, it demonstrated the degree of the commitment of the whole society for recycling. In 2019, a South Korean company, SK Innovation, launched a marketing campaign to educate the public about the different recycling characteristics of the caps and the body parts of plastic bottles, thus they need to be separated when recycling. This initiative was particularly impressive to me because it encompassed the essential elements for successful recycling efforts, such as educational element for the public with detailed knowledge of recycling principles, engagement of the private sector to support government efforts for optimal outcomes, and encouragement for the public to follow best environmental practices.

This comprehensive approach has helped South Korea achieve one of the highest recycling rates in the world, with over 60% of waste being recycled. As a result, the country has significantly reduced landfill usage and minimized the environmental impact of waste disposal.

* Special mention – Despite strong public support and the clearly positive outcomes of its recycling programs, South Korea is now facing new challenges. The rise in single-person households in recent years has been associated with a decline in recycling rates. Additionally, a social phenomenon known as "recycling fatigue" has emerged, with some citizens expressing a sense of weariness over the demands of recycling. Facing these new challenges, the South Korean government may need to adapt its recycling programs to address them effectively.

Local Processing: Once pure materials are collected, the processing efficiency and environmental impacts can be considerably improved. Take plastics for example: Melting pure thermoplastics requires simple melting tanks to turn them into a molten state, then densified forms of pellets, explained in **Chapter II: 10.2 Polymerization (Synthesis) of Synthetic Fibers**. This purely recycled material can be sold at a high value whereas shipping contaminated plastic bottles to overseas facilities, as it is currently done, can cost more than the value of the material, where the former can also lead to a profitable business model for local economies. This model is a part of the solution presented in the **Section 22: Clean Recycling Initiative™** later in this chapter.

***

The comparison of the recycling systems between Montreal and Seoul demonstrates the considerable differences in the environmental outcomes, evidencing strong leadership based on science-based education from governing parties as a key element of successful sustainability efforts.

21.5.2 Textile Recycling

There are both similarities and differences between plastic recycling and textile recycling. Synthetic fibers, which belong to the plastic family, account for almost 65% of worldwide fiber consumption. Fortunately, all these fibers are thermoplastics, meaning they can be "cleanly" recycled using only heat energy. In contrast, there are much higher levels of complexities and impossible recyclability involved in natural and semi-natural fibers. Further details on these subjects will be discussed in the next section.

21.5.3 Other Materials Recycling

Among other recyclable materials, paper, glass, and metals are some of the most commonly recycled. Although recycling rates for these materials vary significantly by region, they generally maintain some of the highest recycling rates.

Paper Recycling: Paper recycling, for instance, benefits from a high degree of process flexibility, allowing it to accommodate relatively larger amounts of impurities compared to the other materials. This flexibility enables the production of such items as egg cartons, moving boxes, roofing underlays, etc., where consumers do not expect an extreme purity. More desirable recycling, however, would be to maintain recyclable paper waste without impurities from foreign materials such as dirt, debris of metals, plastics, glasses, etc., as it will provide more flexibility in converting the materials into higher-value recycled goods and enhance the future recyclability of such materials.

Metal Recycling: Metals, like steel, aluminum, copper, zinc, etc., are highly sought after for recycling due to their particularly high economic values. Similar to thermoplastics, metals can be remelted, reshaped, and reprocessed in closed-loop recycling systems. The higher the purity of the metal waste, the higher the quality of the final product and its continuous recyclability. A negative aspect of the metal recycling is its carbon-intensive nature of remelting at high temperatures. Depending on the types of materials involved, it can easily require several hundreds or even thousands of °C or °F in industrial furnaces.

Glass Recycling: As in the cases of thermoplastics and metals, glass can be remelted and reprocessed primarily with heat energy. Countries such as Germany, Sweden, and Japan have achieved exceptionally high recycling rates for glass, ranging from 80% to 90%, largely due to effective governmental policies and public participation. However, in other countries, recycling rates are significantly lower, and contamination in glass bottles remains as a substantial challenge for more effective recycling efforts. Like the metal recycling, this process requires high heat, typically in mid-1000°C or -2000°F and high carbon output.

***

Clear Advantages of Clean Recycling Initiative™

Reviewing the recycling principles of the four major material types; thermoplastics, metals, glass, and paper, the advantages of the thermoplastics become clear. In the method proposed by the **Clean Recycling Initiative™**, it requires only heat energy, thereby avoiding creating significant quantities of microplastic generation and water usage, typical of conventional plastic recycling. Furthermore, the method does not require any chemical treatment. Numerous other advantages are explained in the following section.

Section 22   Clean Recycling Initiative™

Given my educational and professional background, I sought to leverage my knowledge to create positive environmental impacts in the textile industry and for the public. My motivations were also driven by several other factors, including the widespread greenwashing activities by industry players and the unawareness of elevated toxicity of residual chemicals remaining in textile waste. The later makes textile waste a more imminent threat than most other materials, yet effective efforts have been lacking across all domains, including the industry, the authorities, the NGOs, and the general public.

In the course of developing technologies for my business, HEAT-MX™, I created the technology platform of the **Clean Recycling Initiative™** (CRI). This platform aims to significantly contribute to reclaiming textile materials as well as other types of thermoplastics from waste. Below is the scope of the non-profit technology platform.

Regulated Open-Source Program based on Non-Profit Business Model: The **Clean Recycling Initiative™** (CRI) is a regulated open-source program providing technology solutions for reclaiming textile waste with a minimized environmental footprint. Developing, executing, and maintaining these technologies require significant knowledge and experience in the related fields. Thus, the program is "regulated" based on the existing standards and protocols of the CRI. Industry players and interest groups are invited to participate, and the public can freely use the services it offers.

Origin: As extensively discussed in **Chapter II: Lifecycle Assessment**, the dyeing is typically the most toxic process in textile manufacturing. Residual dyes and other related chemicals remain in textile goods even at disposal, causing long-lasting environmental impacts. However, some textile materials such as thermal insulation do not require dyeing or any other chemical treatment since they are covered by other lining fabrics and are not visible to consumers. My organization, HEAT-MX™, which specializes in the thermal insulation materials and technologies, fully exploited this advantage in developing the CRI technology platform and applying it for the company's products.

Donated for Public Use: Given the importance of this development with potential benefits extending to much wider scopes of textile materials beyond the thermal insulation as well as other industries such as construction, home appliance, automotive, etc., I decided to donate it to the non-profit under the same name of the **Clean Recycling Initiative™**.

Principles of Technology Platform: The most important aspects in the engineering principles of the CRI are:

A. **Zero Water Usage**: The process does not require washing, unlike conventional plastic recycling. This significant difference is due to the collection of clean waste, free of contaminants. Avoiding water usage offers numerous advantages, including zero contamination in water and significant energy savings since drying, typically the most energy-intensive process in textile manufacturing, is unnecessary.

B. **Zero Chemical Use:** The converted materials are used in applications where appearance is not crucial. Examples include noise reduction materials for home appliances and vehicles, energy-efficient curtains for homes and buildings, potential applications as building wraps and thermal breaks in constructions, etc., in addition to the thermal insulation for apparel, outdoor gears, food bags and home textiles. This non-appearance aspect leads to the avoidance of using harmful substances such as dyes and associated chemicals.

C. **Zero Microplastic Creation or Significant Reduction:** The CRI process relies on melting, utilizing the properties of the thermoplastics, and thus does not involve heavy mechanical forces like crushing or grinding. This difference eliminates microplastic creation for the entire process. Some forms of textile waste may require cutting if the mass and density exceed the melting tank's capacity. While cutting produces much less dust than crushing and grinding, it remains an area for improvement – more details provided in the related subjects later in this discussion.

D. **Zero Degradation and Closed Loop of Recycling**: Cleanly processed thermoplastics maintain their physical and chemical properties without degradation, unlike contaminated materials that lose strength due to compromised structural integrity, as discussed in **Chapter I: Section 4 Unfortunate Recycling**. The CRI process ensures the collection and processing of clean textile waste. Contaminated materials are automatically removed from the process to preserve the recycled polymer's structural integrity and enable continuous recyclability.

Intellectual Property (IP): The CRI technology platform is in the patent-pending state as of this writing. Although the initiative operates on non-profit basis and aims to readily provide necessary technologies and services to all applicable parties, protecting the IP is essential to prevent imitations and improper executions. In my extensive industry experience, I have witnessed that copying and imitation for designs, materials, technologies and marketing activities are deeply ingrained practices in the textile businesses. With today's globalized scope, such copying can occur almost instantly.

My concern is that some industry players, motivated to establish their own identity to appeal to consumers, or any other purposes there may be, may attempt to replicate the technology platform without proper knowledge, leading to undesirable outcomes. Therefore, the CRI organization developed the standards and protocols, protected by the IP, and encourage interested parties to contact for necessary technical guidance for implementation.

22.1 Three Levels

The technology platform of the **Clean Recycling Initiative™** (CRI) comprises three distinct levels, each defined by the sources of textile waste and the methodologies for their collection and processing. Under the CRI standards, these levels are designated as LEVEL 1, LEVEL 2, and LEVEL 3.

LEVEL 1　CRI 1

This level pertains to post-consumption waste, which includes waste from households, industries, and specialized applications. Given the current global presence of trillions of pieces of textile goods awaiting the final fate of disposal, it is critical to manage them with minimal environmental impact. The complexity of this process lies in the collection and sorting of waste materials to ensure they are directed to appropriate recycling processes. The main challenge at this level is achieving the highest possible purity in the separated materials, especially since various components such as fabrics, buttons, and zippers are sewn together in the waste. This necessitates carefully designed procedures for collection and sorting, requiring collaboration with the governments of different levels and the companies specializing in material segregation.

Limited Resources: Despite the importance of the LEVEL 1, the CRI organization has not yet been able to allocate sufficient resources to execute necessary tasks. Given the extensive work required for working with municipal and federal governments around the world for establishing the collection and sorting systems as well as the construction and the processing protocols of the required facilities, the organization has put aside the implementation of the LEVEL 1 at present and is, instead, currently concentrating efforts on LEVEL 2 and LEVEL 3 for higher returns under its available resources.

Alternative Options: However, if necessary resources are made available, establishing a system to process large amounts of textile waste from local citizens can yield significant economic benefits alongside environmental advantages. For instance, the city of Montreal can establish multiple sorting and processing facilities to convert large quantities of eligible textile waste into polymer pellets, which could then be sold to manufacturers within or other regions for further processing. As the pellets are in a highly densified form factor, it significantly reduces the carbon footprint for transportation compared with regular plastic or textile waste.

This model can be replicated by any city or municipality globally, positioning them at the forefront of sustainable waste management in addition to economic benefits. It is important to note that a higher efficiency of this effort will be achieved by working together with the brands and manufacturers of textile goods from the initial stages of engineering and development, particularly fabric compositions and construction methods, to increase the purity levels of textile components. This will be further discussed in **22.4 Limitations and Challenges** subject later in this section.

LEVEL 2

The LEVEL 2 pertains to process waste generated in manufacturing environments. For instance, factories producing finished textile goods, often referred to as "cut-and-sew" factories, can generate 5% ~ 30% of incoming fabrics as waste immediately after the cutting processes. The quantity of waste depends on the designs and sizes of the finished goods they produce, and it can even exceed 30%. This process is akin to an individual purchasing a piece of fabric from a local fabric store, cutting it to make, for example, a shirt, and the leftover fabric becoming waste. With the estimated 50 billion pieces of textile goods produced annually, the magnitude of the waste generated from this process is substantial, making it a significant source of the global textile waste. Such process waste is also characterized by the immediate disposal whereas there is a significant time delay between the purchase and disposal of post-consumption waste.

High Efficiency: Recycling in the cut-and-sew environments can be highly efficient. Unlike post-consumption waste, which involves separating various components like fabrics, zippers, buttons, and sewing threads, the fabric waste at this stage of manufacturing can be collected in separate pieces, ensuring desired purity and free from external contaminants.

Flexibility: The LEVEL 2 process can accommodate a wide variety of colors and fabric weights used by manufacturing facilities, thus offer a high level of flexibility in real-life implementation.

Application Examples: A factory, producing various types of textile goods, can adopt the CRI standards in their processes without making any investment. After the cutting process, the factory can sort the waste fabric pieces based on the material types in specific collection and labeling protocols defined by the CRI standards. These assorted fabrics can then be sold and shipped to designated CRI facilities for further processing. If a factory wishes to add more value from their waste, it can make a small investment to install a melting tank and convert the waste into polymer pellets – more to be discussed in the subsequent **Phase 1** discussion. Both approaches allow factories to create economic values from their waste materials, which would otherwise incur disposal costs.

Large Benefits: The same factory, then, can purchase the converted thermal insulation materials, after the entire CRI processes are executed by the CRI processing facilities and its manufacturing partners to convert the materials into thermal insulation, at reduced rates than equivalent materials sold in regular markets. The economic benefits are multi-faceted: Thermal insulation manufacturers such as HEAT-MX™ save on the cost of polymer pellets as they are derived from waste materials, cut-and-sew factories save on disposal costs and gain value from selling waste or polymer pellets to the subsequent processing factories while purchasing the converted materials at lower prices, and finally, brands working with such factories benefit from the lower-cost structure created by the whole CRI process. Consequently, the public enjoys large benefits of the improved environment.

LEVEL 3

The LEVEL 3 pertains to specific types of intermediary raw material manufacturing. For example, HEAT-MX™, the creator and donor of the CRI technology platform, offers a wide range of thermal insulation materials, incorporating this LEVEL 3 technology. During the manufacturing processes of various materials, for example, non-woven sheet products used for thermal insulation, a certain quantity of waste was generated, inherent in the vast majority of the manufacturing processes of this kind. The inevitable waste creation stems from the quality of the sheets on both edges not as consistent as the middle part and is unsatisfactory to meet the strict quality standards of the company. Thus, the edge parts were cut off and discarded, resulting in industrial process waste. Depending on the total width of the non-woven fabric relative to the widths of the edge parts on both sides, this waste can constitute 5 ~ 10% of the input material.

Innovation: To eliminate this waste, HEAT-MX™ adopted an innovative approach, different from other materials in the market. For over four decades, the industry has used a method of blending two different fiber types to achieve specific physical characteristics in a particular category of non-woven materials. However, this fiber mixture makes it virtually impossible to recycle under the scope of the "clean" recycling because the fibers are very small and randomly distributed in the non-woven structure. Separating them would be astronomically expensive, if even feasible.

The HEAT-MX™ organization overcame these recycling challenges by developing a non-woven structure using a single fiber type. As there were clear reasons for mixing the two fibers with one having significantly different physical characteristics from the other, eliminating one fiber type was not an easy task. In particular, the target material for elimination played a critical role in providing a high level of structural integrity for achieving the desired thermal efficiency of the final non-woven sheets.

Nevertheless, persistent efforts, partly facilitated by some unexpected idle time during the pandemic shutdown, led to the creation of the proprietary non-woven structure with 100% homogeneous fiber composition. Immediately following was the implementation of this technology on the company's products and today, the majority of the HEAT-MX's brand customers and manufacturing partners purchase the products incorporating this sustainable technology with no process waste. The products are referred to as "LEVEL 3" in short, and account for staggering 75% of the company's sales revenues in their short existence of slightly over 2 years.

Environmental Benefits: This homogenous fiber composition allows the process waste to return to the beginning of the process and be simply remelted for recycling, using only heat. Furthermore, the edge parts remain within the same manufacturing facility, avoiding the creation of any carbon footprint, which would otherwise be associated with having to transport them out for disposal or recycling. This starkly contrast with plastic waste transported internationally for recycling. The benefits are large and significant.

Cost Benefits: I firmly believe that eco-friendly products should not be significantly more expensive than their conventional counterparts, as this would limit consumer choices. Recent commercial history has demonstrated that while many consumers are willing to pay a modest premium for environmentally friendly products, they are generally unwilling to exceed certain thresholds.

The thermal insulation materials of HEAT-MX™ produced under the LEVEL 3 category of the **Clean Recycling Initiative™** are sold at a considerable discount compared to its competitive materials in the market, offering equivalent or even superior thermal performance. Additionally, these materials provide numerous other performance benefits inherent in the innovations achieved during the development work.

This accomplishment has been made possible through the cost savings from using the materials reclaimed from the edge parts as well as other breakthrough innovations aimed at operational and quality improvements. As quoted earlier, 75% of the total sales revenue of HEAT-MX™ originating from the LEVEL 3 products is a clear proof of the cost-effectiveness, while the company aims to increase the percentage to over 90% within two years as of this writing, or by 2026.

22.2 Execution Phases

The **Clean Recycling Initiative™** (CRI) implementation consists of three distinct phases based on the purity levels of the materials involved: Phase 1, Phase 2, and Phase 3.

Phase 1 - 100% Homogeneous Synthetic Materials

Principles: As previously established, homogeneous synthetic materials such as 100% polyester, 100% polypropylene, 100% acrylic, etc., can be converted into a molten state using heat energy alone. The energy required to melt these materials is significantly lower than that for other recyclable materials such as metals and glass.

For instance, most synthetic materials melt at a temperature range between the mid-100s to low-200s °C (mid-200s to low-400°F), whereas metals and glass require temperatures often reaching thousands of °C or °F. Consequently, the energy consumption for melting thermoplastics is a fraction of that required for operating large industrial furnaces used for metals and glass. Additionally, paper recycling involves significant water usage and energy-intensive processes such as repulping, requiring operating large rotational blades in water and substantial energy associated with it.

Thus, recycling thermoplastics is considerably more energy- and resource-efficient compared to other major recyclable materials. The Phase 1 fully takes advantage of such merits, making it one of the most efficient forms of recycling.

Factory Choices: A factory interested in implementing the CRI's Phase 1 has two options: 1. Sell the fabric waste at lower values to dedicated CRI processing factories or 2. Install simple machinery systems to convert the waste into higher-value polymer pellets. Large factories with substantial waste volumes eligible for the Phase 1 implementation can benefit significantly from installing the machinery systems within their own facilities. Ambitious factories may even integrate additional processes to further enhance the value, though this requires considerably more spaces and costs, complex setups and developing operational and quality expertise in related areas. Currently, the CRI organization focuses on assisting interested parties - brands and manufacturers alike - to either ship their waste materials to CRI processing facilities or convert them on-site into polymer pellets and providing necessary services to design and install the systems that adhere strictly to the CRI standards and protocols.

Simple and Cost-Effective Setup: If a factory wishes to add value by converting their waste into polymer pellets, a simple and cost-effective machinery setup is required within its facility. The core component is a melting tank, which can be customized in various sizes and shapes based on the quantity of the fabric waste the factory creates. For example, a factory generating 5 tons of 100% polyester fabric waste can install a melting tank designed to process this amount.

Fabric pieces are fed into one end of the tank, while polymer pellets are extruded from the other end as the molten polymer solidifies upon exposure to air. This simple setup, costing approximately $70,000 to $100,000 for a 5-ton/day capacity as of this writing, can be offset by the savings from disposal costs and revenues from selling the polymer pellets. As a disclaimer, the CRI organization is not involved in monetary transactions between interested parties such as factories which wish to install the machinery systems and their suppliers, as it only provides necessary services to implement the standards and protocols in place.

Phase 2 - Blends of Synthetic Materials

Materials blended with two or more synthetic materials add complexity due to their varying melt characteristics. For example, PET fibers melt at 250 ~ 260°C (482 ~ 500°F), while polypropylene fibers melt at much lower temperatures, between 160 ~ 170°C (320 ~ 338°F). Therefore, when these two materials are blended in fabrics, the melting temperature in the tank must be raised close to or above that of the higher of the two, the PET in this case, before the subsequent extrusion process.

Melt Flow Characteristics (MFC) and Inconsistency: In a melting tank containing such combinations, the PET polymer remains in a more viscous state, whereas the polypropylene has a considerably lower viscosity at a desired processing temperature of the tank. The viscosity is generally described as the Melt Flow Characteristics (MFC) or Melt Flow Index (MFI). While the MFC of 100% homogeneous material is consistent and easy to control during extrusion, blended materials present inconsistency. This inconsistency limits the outcomes and form factors of the extruded polymers: Consistent MFC allows for the extrusion of thin fiber form factors. However, inconsistent MFC prevents fiber formation because the low-viscosity polymer behaves like water, whereas the high-viscosity polymer flows like honey or heavy motor oil.

Factory Choices: Due to the limitations, the Phase 2 involves converting the molten polymers into sheet formations, similar to those of film making or molded form factors. These sheets can be used in applications such as building wraps, strength reinforcement materials, etc. Various chemical or mechanical means can be incorporated inside the melting tank or during the extrusion process to provide specific characteristics to the converted sheet. Intended usage for the molded plastics from such process must be able to take account for inherent structural weakness on its own in comparison with homogeneous materials.

The machinery setup for the Phase 2 can be more complex. It may involve a specialized extrusion process converting the molten polymer into a sheet or other form factors desired for specific subsequent processes and final products. This setup must be able to accommodate the variations in MFC and adjust the resulting form factors of the output materials accordingly. More technical details can be obtained by contacting the CRI organization.

Limitations: Highly blended fabrics with multiple fiber types with widely different melting characteristics present significant challenges in setting operational parameters, although not impossible. The CRI organization can work together with the parties facing such challenges for more desirable outcomes.

Phase 3 - Blends of Synthetic and Natural Fibers

Blended materials containing both synthetic and natural fibers utilize the principle of melting synthetic fibers while retaining natural fibers in their original forms. This is because natural fibers, when subjected to heat energy, do not melt but instead burn.

Upon heating blended materials in a melting tank, synthetic fibers will melt while natural fibers will remain suspended in the melt. The Melt Flow Characteristics (MFC) of such blends vary widely depending on the compositions of the input materials. Generally, a higher proportion of synthetic fibers results in more favorable extrusion conditions.

Limitations: A lower proportion of synthetic fibers does not provide sufficient liquidity to achieve the desired characteristics of the melt solution necessary for extrusion. Furthermore, the extruded materials will exhibit the characteristics of both synthetic and natural fibers, where higher proportion of natural fibers increases chemical reactivity and degradation rates while reducing strength. Therefore, understanding the melt characteristics of the blended materials in relation with the input material compositions is crucial for the effective implementation of the Phase 3. More technical details can be obtained by contacting the CRI organization.

22.3 Memberships

There are currently two types of memberships: Individual and Corporate Membership while a new category of Educational Membership will be added.

Individual Membership: The Individual Membership is completely free to join and entails no monetary obligations for annual maintenance. Members benefit from receiving woven labels designed to be attached to their textile goods before disposal. These labels aid sorting facilities in segmenting and separating the fabrics and other materials into designated categories. Members will only be charged for postage.

This aspect pertains to the LEVEL 1 of the CRI platform, which is not currently implemented as of this writing. Once the required systems, including collection, sorting, and processing, are established for various municipalities, the members in the applicable regions will be notified. Interested individuals can join the free membership through the provided link –

https://cleanrecyclinginitiative.com/membership/

Additionally, readers can understand the scope of the CRI organization's efforts or contact the organization for general inquiries from the same website.

Corporate Membership: The Corporate Membership is designed for the organizations that benefit from the CRI platform's technologies and services for commercial purposes. Benefits of becoming a member company include necessary services for implementing the systems mentioned in the LEVEL 1, 2, and 3 and the Phase 1, 2 and 3. Furthermore, other engineering services which relate to enhancing the recyclability of the member companies' merchandises are offered.

The Corporate Members pay basic membership dues, which support the CRI operations based on the non-profit model. Unlike many other certification programs that involve significant costs for initial certification and ongoing maintenance, the CRI Corporate Membership program is structured to reflect efficient operations, resulting in affordable rates for the member companies. The fee structure is scaled to accommodate businesses of various sizes, with minimal fees for small companies and relatively higher fees for medium and large companies. The whole system is created to easily facilitate all interested parties with no financial burden regardless of their sizes.

For the implementation of the technologies in manufacturing facilities, where CRI personnel may need to travel to assist, associated travel and administrative costs may be charged based on the scope of each project. More information is available by contacting the organization.

Finally, while it is not implemented at the time of writing, the Educational Membership category will soon be created and offered for free. With a proof of educational operations and the identification of target students, all applications regardless of the ages or any other background profiles of the students will be accepted to gain the free memberships.

22.4 Limitations and Challenges

Material Engineering: Given the extensive diversity in the compositions of the fabrics created by numerous industry players including brands and manufacturers, recycling them with minimal environmental impact presents challenges.

However, certain considerations and actions during the stages of product engineering and development can enhance the recyclability with higher value creations. As demonstrated by HEAT-MX™'s innovation in modifying the fiber compositions, deviating from 40 years of the industry practices in its non-woven sheet formations, opportunities exist to make fabrics more favourable for optimal recycling. While some fabrics, such as woven fabrics with elastic yarns for high stretchability, may not have alternative choices to match the same level of stretch performance without using required elastic yarns, many blends of yarns and fabrics can be re-engineered for better recyclability with either no or little compromising of performance.

Engineers of textile brands and manufacturers, therefore, can have a close examination at the existing blends they use and seek for opportunities to improve the recyclability of each fabric.

Potential Microplastic Generation during Cutting: In the engineering of the melting machinery, the feeder for the input fabric materials into the melting tank can lead to requiring a cutting mechanism. In the vast majority of factories where fabric pieces are small, the design of the melting tank can be engineered to easily accommodate the fabric pieces without such cutting need and convert them to a molten state. In such cases, there is no concern on microplastic generation as the method relies solely on the melting characteristics of the involved polymers.

However, if a factory produces large and heavy pieces of waste and put them into the tank in large quantities, the fabric feeding may exceed the tank's capacity to melt them in a given process condition and time frame. In such cases, cutting the fabric pieces may be necessary. Although cutting with sharp knives generates much fewer microplastics compared to the crushing and grinding used in the conventional plastic recycling, it still produces some dust. Careful design of the feeding and melting system can minimize the need for cutting, but if required, a suction system in conjunction with the cutting to capture the dust and direct them into the melting part of the tank can eliminate such concern. CRI's technical services can provide necessary guidance on implementing these measures.

Energy Use: As mentioned earlier, converting thermoplastics into a molten state requires heat energy although it is significantly less than the energy needed for melting metals and glass. While this is advantageous in every aspect of material recycling, it is still an energy-consuming process. To maintain the transparency and fairness throughout the book, I did not wish to omit the acknowledgement that the CRI process uses energy, albeit in the applicable contexts and much smaller amounts.

Risks of Exploitation: The risk of exploitation has always existed in the textile industry, whether related to designs, technologies, or materials. Any party misusing good technologies can lead to negative outcomes. To mitigate these risks, the CRI organization operates under the strict standards and protocols to ensure proper use and environmental protection.

***

Industry experts who are familiar with this **Clean Recycling Initiative**™ platform considers it as one of the most important technological advancements since the invention of the synthetic fibers as it provides an optimally designed recycling structure on a tremendous amount of textile waste, particularly synthetic. While some key programs have already been implemented in specific levels and phases in different manufacturing settings, there are numerous other areas to spread the benefits. Considering every household, textile brand, manufacturer and other directly and indirectly associated players in the industry can benefit from it, a tremendous undertaking belies ahead of the **Clean Recycling Initiative**™ (Non-Profit) organization.

Collaborative efforts among the parties of Interest, including industry players, NGOs, government organizations, etc., can further increase the efficiency and benefits of the program.

Section 23 BE GREEN HEROES (BGH)

The non-profit organization, **Clean Recycling Initiative**™ (CRI), launched the **BE GREEN HEROES** (BGH) campaign with a clearly defined objective of the fight against misinformation. The campaign considers this objective as the very first and most important step towards more effective global efforts for the environment and sustainability.

Dr. Clair Patterson

The term, "GREEN HERO", was inspired by the remarkable work of Dr. Clair Patterson (1922 ~ 1995). During his research on calculating the age of the Earth using a specific lead isotope, Dr. Patterson encountered persistent errors due to significant lead contamination in the atmosphere. He soon realized that the levels of the atmospheric lead exceeded natural occurrences, indicating an external source of contamination. His investigative work revealed that petroleum companies were adding lead to fuel to enhance the engine performance of vehicles.

Fight against Giants: Dr. Patterson recognized the severe health implications of the lead pollution and shared his findings to advocate for the cessation of the lead additives in fuel. At that time, petroleum companies were among the largest corporate entities before the likes of Apple, Google, Nvidia and Amazon of today, and possessed dominant and vast socioeconomic influences. Driven by their operational interests, the companies rejected Dr. Patterson's work. Despite facing substantial opposition from these powerful stakeholders, Dr. Patterson remained steadfast in his efforts.

Persistence and Dedication for Humanity: Through Dr. Patterson's relentless work, the leaded fuel was eventually banned in the United States in the 1990s, two decades after he first raised the alarm. His heroism lies not only in his scientific acumen to understand such pollution but also in his unwavering dedication to protecting the humanity from such toxic pollutant in the air. Although not as widely recognized as Albert Einstein, Dr. Patterson's contributions have profoundly impacted our lives by improving the quality of living for everyone on earth, earning my utmost respect as GREEN HERO.

BE GREEN HEROES (BGH) Campaign

The BGH campaign encourages the public to become informed with accurate science related to environmental sustainability and spread their knowledge to others to prevent further harm caused by misinformed or ill-conceived claims and actions.

The campaign does not expect individuals to suddenly develop expertise in chemistry or related fields like Dr. Patterson but rather to advocate for informed and responsible actions for the environment. For instance, the knowledge readers gain from this book can be a good conversation piece in a dinner gathering with friends and family. It can be of textile related subjects such as the greenwashing examples discussed in the earlier chapter or the concerns over the EV conversion. Any effort to raise awareness or inform others with accurate science against notion-based unscientific claims and actions would be a good act of the GREEN HEROES.

Furthermore, the BGH campaign provides detailed action plans and insights for effectively fighting against misinformation. A series of the following discussions highlight the details:

23.1 Question Everything

Do not equate the authority or the size of organizations with information accuracy. Many large organizations, including private enterprises seeking profits, NGOs, government-associated entities, and even government organizations themselves, have made numerous erroneous claims, engaged in misguided actions, and participated in greenwashing. Instead of developing necessary expertise, these organizations often opt for easier paths, exposing themselves to the influences of self-proclaimed experts who lack requisite experience and scientific knowledge in relevant fields. This is one of the most significant issues facing the sustainability landscape today.

23.2 Check Credentials

A paper recently published claimed that "*Plastics have never been recyclable and never will be*," challenging the foundational principles of the polymer science. What was particularly surprising was that this paper was published by a prominent think-tank based in Washington D.C., with the first author holding a PhD degree. The paper's attributes, including the prestige of the publishing organization, the academic title of the first author, and even the location where many influential associations in related fields, such as American Chemical Society (ACS) and American Petroleum Institute (API), are located, could easily lend undue credibility to its content for the general public, many of whom may lack sufficient knowledge in related areas to critically assess the inaccuracies presented.

PhD in History: Upon reading the paper, I was struck by the level of scientific inaccuracies it contained, prompting me to investigate the background of the first author. It turned out that the author held a PhD in history. While I do not undermine interdisciplinary expertise, exemplified earlier with the book, "*Material World*", written by a journalist, Ed Conway, it was clear that the PhD author did not possess such high quality as Ed, yet made many unscientific and notion-based claims in the paper.

A common issue in the realm of environmental and sustainability discussions today is that many people tend to accept information at face value without scrutinizing the qualifications or expertise of the authors. This leads to widespread propagation of misinformation across the internet and in social interactions.

23.3 Consult with Real Experts

As discussed in **Chapter IV: Section 18 Organisations Lacking Expertise**, identifying genuine experts combining experiences in manufacturing and real-world science within the Western world is challenging. Instead, the landscape is often populated by self-proclaimed experts who are readily accessible and prominent in various environmental and sustainability-related activities, such as internet presence, international conferences, podcasts, etc.

Importance of Bridging Knowledge Gaps: Despite the difficulty in finding true experts, it is imperative to seek them out. Some forward-thinking companies are already addressing this gap by bringing employees from manufacturing hubs in Asia, South America, etc., to their headquarters. This practice facilitates the transfer of real-world knowledge and experience to the Western context, enabling a more balanced perspective that incorporates insights from both sides of the world. These individuals can provide valuable insights that are grounded in reality, rather than theoretical or superficial understandings. Integrating their knowledge into sustainability efforts will lead to more effective and realistic solutions.

23.4 True Experts - Let Your Voices Heard!

As repeatedly emphasized, the voices from self-proclaimed experts are excessively dominant. The quality of hands-on experiences combined with academic backgrounds is essential to fight against this undesirable force and make impactful decisions for the environment and sustainability.

Many true experts in the real-world environments of related industrial activities are found in manufacturing hubs and English may not be their first language. Unfortunately, our current social structures often exclude these voices from global sustainability efforts, despite their invaluable insights. To address this, we must actively seek and integrate their perspectives; the work which can be coordinated by a global organization such as the currently fictitious "Nations Unified for Environment and Sustainability (NUES)", discussed earlier.

23.5 Be Informed and Educate Others – BE GREEN HEROES!

GREEN HEROES are encouraged to stay informed with accurate, science-based information and to actively combat misinformation wherever and whenever possible. To address the ongoing crises in sustainability and ESG (Environmental, Social, and Governance), it is critical that the GREEN HEROES voice their views to raise awareness and spread accurate information.

We live in an era where significant societal efforts are dedicated to environmental sustainability. Interest levels in environmental management and ESG have never been higher. However, if misinformation continues to be the basis for our collective understanding, it can lead to detrimental and irreversible outcomes. It is a critical time for the GREEN HEROES to act!

Chapter V Post-Chapter Commentary

Like everyone else, my knowledge and experiences are inherently limited. I recognize that I am constrained by my own perspectives and preconceived notions just like everyone else. It is inevitable that I may be mistaken in certain areas or overlook important points, subjects, and concerns in this book. Undoubtedly, I will continue to reflect on what I could have done differently or included, even long after this book is published.

However, one of the key messages I attempt to convey in this book is abundantly clear. Though it may seem harsh, I sincerely hope that those who are not genuine experts take the time to acquire necessary knowledge before disseminating information in public domains. As I opined earlier, information accuracy is a critical element for our society to make effective efforts for environment.

Urgency: Although this book could have encompassed more areas and subjects or been more refined, I decided to publish it promptly to address several key points urgently. My hope is that the information and perspectives presented here will provoke thoughts for many and stimulate debates and discussions, ultimately promoting deeper engagement from the public with these critical subjects and issues.

Broader Scope for Humanity: The world is rife with discrepancies, and sustainability efforts are no exception. As highlighted in the discussion on the EV conversion, it is truly inappropriate that so many miners in underdeveloped and developing countries must endure extremely poor working conditions while companies in the EV industry enjoy billions of dollars in profit, benefiting executives with millions. Sensible allocation of even a fraction of this wealth could ensure proper safety equipment for the artisanal miners. While this book is not about economics and wealth distribution, it is of my opinion that this inconceivable level of discrepancy will negatively impact the collective efforts we make in pursuit of humanity's sustainability.

***

In this chapter, I am akin to David challenging Goliath, questioning such corporate behemoths as Tesla, BYD, Google, etc., and even some of the most influential authorities like the UN, the EU Commission, the WHO, the US EPA, etc. In my justification, it is not merely about challenging these giants but about speaking the truth. I commend myself for having the courage to address these issues for the sake of the humanity's sustainability. I hope that readers of this book will also find courage to act. While the general public may not be aware of the critical state we are in, you now have been informed and have a clearer path on the next step to take:

BE GREEN HEROES and inspire friends and family to do the same!

Closing Thoughts

Fear Factor

Our lives are filled with diverse emotions, from positive feelings like joy and hope to negative ones like fear and guilt. Unfortunately, negative emotions often have a more profound impact on us. While feelings of overjoy don't usually require medical attention, anxiety might lead us to seek treatment. This explains why bad news travels faster and engages us more deeply, particularly through the fear of unknown futures.

Nature of Predictions: This phenomenon is evident in the collective social mindset that shapes our behaviors and reactions to predictions. Often, such predictions provoke fear and deeply influence our perceptions. A historical example is the widespread interpretation of Nostradamus's cryptic prophecies as foretelling the world's end at the close of the last millennium, which became common public knowledge for centuries and stirred significant concern for many.

More directly related to the theme of this book, I remember the predictions made during the 1970s and 1980s that warned of imminent oil shortages. These forecasts suggested that oil reserves would be depleted within 30 years, implying that by the early 2010s, the world would face an energy crisis that could plunge us back into a primitive way of life. These dire predictions were put forth by individuals considered experts at the time, such as industry engineers, PhDs, professors, and seasoned journalists who presented seemingly reliable scientific data such as current reserve levels and rising consumption rates to support their claims of looming depletion.

However, this disastrous outcome has not occurred. In fact, today we have more oil reserves than we did in the 1970s and 1980s, thanks to advanced technologies that have discovered new deposits. Setting aside the on-going environmental debates, as it is not the topic of this particular discussion, the recent shale gas revolution, utilizing hydraulic fracturing and horizontal drilling, has enabled extraction in areas previously not considered as reserves. As a result, many experts now estimate that we have more oil reserves than we will ever need as the world rapidly shifts towards renewable and nuclear energy sources.

Nature of Threats: It goes without saying that we must do everything in our power to address the challenges of climate change. The scope of its consequences can be overwhelming, and while there are countless concerns, individuals tend to have varying degrees of concerns on different issues. Some are more alarmed by certain phenomena, while others are more influenced by different aspects. This is only natural, given the vast number of environmental problems we face today. When it comes to carbon impacts, my greatest concern lies with the phenomena occurring in formerly permafrost regions, where methane (CH_4) gases are erupting from the ground, bubbling up in lakes, and ancient pathogens are being revived.

In the meantime, the perspectives I have presented in this book relate more to adequately addressing the causes which can create more immediate impacts from cumulative effects, corresponding to the toxicity theme presented throughout the book. It is akin to people caring about washing off pesticides thoroughly on ingredients while keeping the stove on only for cooking as both need to be done for healthy living and environmental benefit.

The science is clear: cumulative effects from both reactive and inert chemicals dictate whether the consequences are more immediate or gradual. While climate change has taken the center stage of global environmental efforts, the broader implications of these chemically reactive substances must not be overlooked.

Overcoming Challenges

Although I have addressed many serious concerns throughout this book, the reality is that the world is not coming to end any time soon whether by the causes and views expressed by some climate extremists or the toxicity concerns I have raised.

Reflecting on the history of Earth and humanity, we understand that the most likely causes of catastrophic events remain large asteroid or comet strikes, massive volcanic eruptions, or devastating microbial outbreaks. From a climate perspective, science reveals that cooling, rather than warming, has historically caused greater damage. For instance, the Ordovician-Silurian Extinction (~444 million years ago) saw the disappearance of approximately 85% of marine species, largely due to global cooling.

I acknowledge that this statement might spark debate, especially in the context of ongoing discussions about climate change. However, this book does not aim to take sides in the debate over the validity or extent of the ongoing climate change's impacts. Instead, I encourage readers to approach the topic with a balanced perspective, as there is often more to the story than what is presented. While climate activism dominates public discourse, it is essential to consider insights from different viewpoints. With that in mind, I recommend two books that offer important perspectives on environmental efforts:

* **Not the End of the World**: *How We Can Be the First Generation to Build a Sustainable Planet by Hannah Ritchie*

* **False Alarm**: *How Climate Change Panic Costs Us Trillions, Hurts the Poor, and Fails to Fix the Planet by Bjorn Lomborg*

Naturally, there are many parts of the books I agree with and others I do not. However, an important message of these books, which I also strive to convey herein, is that we must avoid panic. We are not doomed.

The Dutch have successfully managed rising sea levels since the 14th century, and humanity has endured changing weather conditions since prehistoric times. Early humans, such as Homo Erectus, survived multiple ice ages without modern luxuries like insulated homes or advanced winter gear. Humanity's resilience and adaptability have allowed us to thrive despite numerous climate challenges.

This adaptability extends also to the way we address current environmental challenges from a different angle. Poor human decisions in many industrial activities have significantly impacted the environment and our sustainability. Actions based on comprehensive assessments and effective efforts can lead to starkly different outcomes than those based on narrow views and unscientific notions. Focusing only on climate change while ignoring toxicity is like caring for one's skin while neglecting the health of internal organs. Or, concerning only on energy use in cooking while omitting to thoroughly rinse pesticides and other harmful chemicals in vegetables would be analogous to such approach.

Today, we have many influential organizations such as the World Meteorological Organization (WMO), various UN departments, local and federal governments of different countries around the world, etc., actively promoting awareness and actions on climate change. However, no equivalent organization like a "World Toxicity Organization" exists to address potentially more harmful toxicity concerns comprehensively. Instead, this aspect is often downplayed by other influential bodies like the World Health Organization (WHO) and the Environmental Protection Agency (EPA), which employ irresponsible practices such as setting "allowable" or "tolerable" intake limits for harmful substances. This approach does not adequately protect humanity from these substances that accumulate and persist in nature.

Many governments have set goals through the Paris Agreement, assuming continuous human existence: carbon goals for 2030, long-term targets for 2050, and net-zero emissions between 2050 and 2100. However, the nature of toxicity accumulation challenges this assumption. Analyzing the two deadliest events in human history outside the mass extinctions from catastrophic asteroid impacts and volcanic events addresses a related context: The Black Death (Bubonic Plague, 1347-1351), with an estimated death toll of 75-200 million, caused by the bacterium Yersinia pestis, spread through fleas on rats and devastated Europe, Asia, and North Africa and the Spanish Flu (1918-1919), with an estimated death toll of 50-100 million, resulted from interactions between human body and deadly microorganisms. Harmful chemicals share similar principles of chemically reacting in our bodies and can create much more immediate and detrimental impacts than changing weather conditions.

Need for Actions: This highlights the need to address toxicity in our environmental management. The work necessary to effectively deal with the concern involves essential education based on accurate science with comprehensive scopes.

Human evolution follows a clear path of continuous population growth and increased environmental impact. Ever more electronics and EVs, containing highly reactive chemical substances and consequential environmental footprints, are sold every minute globally. Environmentally harmful mining and textile production will continue to increase their outputs. Unfortunately, organizations with significant influences in our lives have catastrophically failed in effectively addressing these threatening environmental issues. As a result, there is a glaring absence of leadership to steer humanity in the right direction.

While finding ways to alleviate the toxicity burden from the ongoing human living is of critical importance, we must also recognize that there are no shortcuts or simple solutions. The challenges we face are immense, and addressing them will require a slow, deliberate process coupled with highly effective efforts. In my view, the path to a healthier environment and genuine sustainability begins with tackling the widespread misinformation that impedes progress. By empowering the readers of this book, we can inspire them to become **GREEN HEROES** in the fight for a better future.

***

As I publish this book, many thoughts come to my mind. Given dominant perceptions existing in our society and starkly contradictory views I presented herein, I anticipate varying reactions from readers. Despite my best efforts to provide related information grounded in science and practical experiences, I know skepticism will persist among many readers.

While I don't compare my work to history's greatest scientists, it is worth noting that Nicolaus Copernicus' heliocentric theory - that the Sun, not the Earth, is the universe's center - faced significant scrutiny. Throughout history, however, great minds have inspired others to seek truth, leading to scientific advancements and bettering our lives. Conversely, Copernicus' ideas influenced Galileo Galilei, who improved the capability of telescope technology and observed clear evidence of Earth's roundness and movement around the sun. However, Galileo faced prosecution and house arrest upon revealing the evidence despite his effective efforts.

Similarly, I anticipate significant challenges following this book's publication, albeit no house arrest :) Some may read the book skeptically from beginning to end. Others may begin with doubts and turn them into denial. Just as we still encounter flat-earthers and conspiracy theorists who reject the most convincing scientific evidence and practical experiences, such as airplanes flying over the globe and cosmic photos of Earth, it is inevitable to face similar challenges in these discussions. However, I hope that most readers evaluate the contents with open minds and, at the very least, question their priorly existing perceptions on the topics presented in the book.

In my view, these questions and doubts are healthy phenomenon. I often tell people around me not to blindly trust my words. If they care about the environment, they should invest time in research to form their own convictions rather than relying on me or any other information sources for that matter.

Unfortunately, however, this is not an easy path. Achieving adequate education in the current state of global environmental management requires significant and effective effort, as misinformation is prevalent in our society today. It is far easier to be influenced by and echo existing perspectives than to seek truth, and many people may not voluntarily make such efforts in their free time.

While I discussed the negative outcomes of readily available misinformation on the internet, an analytical mindset can navigate this landscape to seek necessary validity from various sources. This involves earnest research with a comprehensive scope, including academic papers and reliable industry sources. If there is no evidence supporting certain claims or actions, no matter how basic they may seem, such information should not be accepted at face value.

For most readers, engaging in such thorough analysis may not be feasible. Emphasizing once again, gaining comprehensive perspectives on environmental sustainability is not about echoing random information or adopting general notions without scientific bases. Hence,

the final words of the book are dedicated for those without scientific background or time to do such investigative work:

This book offers nearly 700 discussion points designed to help readers develop skills and analytical mindsets for evaluating information, claims, and actions in the environmental and sustainability realm. Furthermore, I made several suggestions to be informed with accurate science and fight against misinformation in **Chapter V: Solutions**, which will guide readers in the right directions. Finally, you can join the "**BE GREEN HEROES (BGH)**" movement by following the **Clean Recycling Initiative™** (Non-Profit) on various social media platforms to stay informed on various topics. It is a critical time for the GREEN HEROES to act.

***

LinkedIn:

https://www.linkedin.com/company/78306454/admin/dashboard/

or

https://www.linkedin.com/groups/14412231/

Facebook:

https://www.facebook.com/cleanrecyclinginitiative

Instagram:

https://www.instagram.com/cleanrecyclinginitiative/

***

From the Author

Thank you for reading! If you enjoyed this book, I would love to hear your thoughts. Your reviews help other readers become GREEN HEROES and support the cause. Simply click the link or scan the QR code below to leave a review on Amazon. Your feedback means the world and I read every review!

https://www.amazon.com/GREEN-HEROES-MISINFORMATION-ENVIRONMENTAL-SUSTAINABILITY-ebook/dp/B0DHLRFTVW/ref=tmm_kin_swatch_0?_encoding=UTF8&dib_tag=se&dib=eyJ2IjoiMSJ9.WVq8Wn4IX2kFgaRZF3VEiw.4HK_LkYil6AL_VbDTm-gLpjLFcIMEk0UGSlOprTqRgQ&qid=1754485455&sr=8-1

Finally, this book is accompanied by a companion resource designed to deepen understanding of textile products, with a particular emphasis on the most material- and chemical-intensive segment: winter textiles. You can access it via the link and QR code provided below.

COLD WINTER WARM WINTER: Same Weather Different Protection – A Smart Guide to Staying Warm from Everyday Life to Advanced Textile Innovations, available globally via Amazon, including the following link and QR code:

https://www.amazon.com/COLD-WINTER-WARM-PROTECTION-INNOVATIONS/dp/B0FJY5464M/ref=tmm_pap_swatch_0?_encoding=UTF8&dib_tag=se&dib=eyJ2IjoiMSJ9.OIliAeUlbMDZ8WTV9oq_fQ.JvAvlMTMFPw0clYUzl1T0P2lPHXUBYmRXeCcQ84wdXw&qid=1753916438&sr=8-1

www.ingramcontent.com/pod-product-compliance
Lightning Source LLC
Chambersburg PA
CBHW052235220526
45471CB00001B/47